heating with wood

Andy Reynolds

L I L I

Published in April 2008 by

Low-Impact Living Initiative
Redfield Community,
Winslow, Bucks, MK18 3LZ, UK
+44 (0)1296 714184

lili@lowimpact.org
www.lowimpact.org

ISBN 978-0-9549171-5-9

Illustrations: Mike Hammer
Picture acknowledgements
All photographs by Andy Reynolds apart from
Author photograph, page 9, Pete Bosshard
Chainsaw illustrations, pages 134, 135, 136, Lantra Awards
Cell Structure, page 39, Macmillan Publishing

Printed in Great Britain by
Lightning Source, Milton Keynes

contents

illustrations 7
about the author 9
introduction.. 11
background.. 13
    some social history 13
    carbon.. 14
    carbon footprint... 15
    plant chemistry 15
    how wood burns 17
    sustainable timber production.................................... 18
buying and storing wood 21
    different methods of buying wood............................... 21
    cutting your own wood... 26
    chainsaw training... 27
    storing firewood 28
    more about wood burning.. 30
    firewood processing... 30
    hand saw 31
    reciprocating saw... 31
    chainsaw.. 32
    saw bench 32
    firewood processor 33
    splitting 33
    kindling 35
how timber dries 37
    wood structure 37
    moisture content.. 38
    cell structure and drying 40
    bark.. 41
    annual ring structure... 41
    species 43
growing your own fuel 45
    grants.. 46
    planting 46
    weeding 47
    species choice 47
    harvesting 48
    managing existing woodland 49

charcoal 51
   charcoal production..51
   setting up a burn 52
stove design..57
   stove door 57
   primary air intake...57
   secondary intake...57
   chimney damper 57
   baffle 57
   chimney...58
   ash grate 58
   log grate 58
   back boiler..59
   types of stove..63
   box stove..63
   front loading stove..64
   Clearview stoves 65
chimneys..67
   chimney deposits 68
   chimney design 68
   new building 70
   retrofit...72
   downdraught 73
   wind...74
   liners 75
   smell...75
   chimney sweeping 76
water heating systems 77
   hot water tank 79
   heat link...80
   central heating 81
   solar water heating...83
automated systems...85
   wood chip..85
   pellet 86
   feed systems...88
   hoppers 89
   combustion systems..89
   multi-fuel boilers..90
   batch-burner boilers 91
   fuel supply..94

stove size.. 94
district heating 95
heat store... 95
what system is best for me? 95
beginner's guide to selecting and installing a stove 97
cost... 97
heat output... 98
design 98
second-hand stoves 99
installing a stove 100
beginner's guide to stove use... 103
preparation 103
clean air act 108
cooking with wood 111
cooking heat 114
regulating oven temperature... 116
home stove building 117
door and frame 117
hinges 121
making a round stove 122
air inlets 123
building regulations... 124
sharp bits.. 125
personal protective equipment 125
hand tools 127
small axe 127
splitting axe and hand saw 129
sledge hammer and wedges 130
saw bench 131
chainsaw... 131
electric chainsaws 131
petrol chainsaws.. 132
chainsaw safety features 132
fuel and oil 134
chain oil 135
bar and chain... 135
kickback.. 136
chain brake 136
viewpoint 137
pitfalls 137
resources.. 141

wood stove suppliers / installers 141
firewood / kindling suppliers 143
courses / training... 144
books 145
information / associations 146
forestry / trees 148
tree nurseries 149
other... 151

illustrations

fig. 1: carbon/oxygen cycle for growing and burning wood 16
fig. 2: forestry contractor hard at work 24
fig. 3: timber lorry unloading firewood 25
fig. 4: slab wood in the firewood stacks 26
fig. 5: wood sheds with plenty of ventilation for drying. 29
fig. 6: reciprocating saw 31
fig. 7: tractor driven saw bench (guard removed for clarity) 33
fig. 8: different types of axe 34
fig. 9: chopping block at a suitable height for a wheel barrow.... 36
fig. 10: hardwood and softwood structure 37
fig. 11: cell structure 39
fig. 12: ash that has grown quickly showing annual rings 42
fig. 13: charcoal kiln parts 52
fig. 14: charcoal burn in progress 54
fig. 15: fireplace conversion with door open 60
fig. 16: fireplace conversion showing door hinges 61
fig. 17: fireplace conversion with door closed 62
fig. 18: box stove 64
fig. 19: Clearview air inlet 65
fig. 20: Clearview air inlet (2) 66
fig. 21: register plate & soot trap 69
fig. 22: a stainless steel, double-wall insulated flue 71
fig. 23: smoke spiral 73
fig. 24: simple hot water system 78
fig. 25: heat link 80
fig. 26: central heating 82
fig. 27: wood fuel and solar hot water connections 84
fig. 28: wood chip 86
fig. 29: wood pellets 87
fig. 30: feed auger and agitator 88
fig. 31: biomass flames 89
fig. 32: 70Kw Alcon biomass boiler 90
fig. 33: basic batch boiler 92
fig. 34: Alcon 30kw biomass batch burner 93
fig. 35: paper twists 104
fig. 36: kindling 105
fig. 37: lighting a fire 106
fig. 38: kindling burning 107

fig. 39: stove lit and doors closed 108
fig. 40: Rayburn no. 3 112
fig. 41: Rayburn gas path..113
fig. 42: Tirolia gas path 114
fig. 43: Tirolia Cooker..115
fig. 44: door frame..118
fig. 45: door carcass 118
fig. 46: doorframe, door and seal...119
fig. 47: door seal 120
fig. 48: door seal groove 121
fig. 49: stove door and hinges..122
fig. 50: round stove 123
fig. 51: chainsaw PPE 126
fig. 52: preparing kindling (1) 128
fig. 53: preparing kindling (2) 129
fig. 54: sledge and wedge...130
fig. 55: chainsaw safety features 133
fig. 56: saw chain parts 135
fig. 57: chainsaw cutter..136

about the author

Andy Reynolds moved to the Lincolnshire fens in the early 1980s, with his partner Geri Clarke, and renovated a cottage that had had very little work done on it since it was built in 1840; heating with wood was seen as a cost-saving venture and a way of heating the home from the surrounding resources. The heating system they installed was fitted with recycled radiators and a large wood stove. This was when Andy's real experience of wood procurement, processing and seasoning began. He soon learnt that there is nothing so bad, in the wood-fuel world, as a stack of green and unseasoned firewood with nothing available that's suitable to burn at the moment.

Andy is a carpenter and joiner and once ran a joinery company that specialised in staircases and through this he gained knowledge of how wood dries by using air- and kiln-drying of locally-sourced timber.

As Andy's awareness of local timber resources and timber quality grew, the time seemed right to plant three acres of their smallholding with broadleaved tree species, with the hope that, in the twenty-first century, some useful sawlogs would be produced.

This interest in growing trees for timber, and the related environmental and habitat values, eventually led Andy along a path to a degree in Forestry Management: an educational route that had life-changing, positive results. This slowly-evolving path, that included qualifications in teaching and as an instructor, has not only resulted in Andy teaching subjects he is passionate about, but also the writing of this book. Andy currently runs the

Woodland Training Division of the International Business School, see 'resources' (page 141), which specialises in chainsaw, arboricultural and forestry training, with his partner. He also provides forestry management consultancy for several estates in Lincolnshire, helps to run the Ecolodge, see 'resources' (page 141) on the smallholding, and has occasional expeditions to regions out of the county to teach for the Low Impact Living Initiative (LILI).

The Ecolodge was built from locally-grown timber and as many recycled materials as possible. The idea, conceived in the late 1990s, was not only to reduce the impact of the building on the environment, but also to provide a quiet holiday experience that introduces guests to wind and solar electricity, the use of rainwater for everything except drinking, a wood-burning cooker, as described in 'cooking with wood' (see page 111), and a compost toilet. Through the Ecolodge he and his partner hope to encourage an awareness of how things could be different, and more enjoyable, by reducing consumerism and the day-to-day wants to a minimum.

introduction

Welcome to *heating with wood*, a book that is intended for anyone thinking of installing a wood stove and associated heating system, or who wants to know more about this almost carbon-neutral form of space and water heating.

It was commissioned by LILI following several successful *heating with wood* residential courses and reflects the nature of those courses with its detailed approach to a wide spectrum of relevant subjects and by encouraging practical activity and hands-on experimentation.

There are several chapters covering the practicalities of wood as a fuel: including everything from where and when to buy and how to store firewood and how much you will need, to growing your own fuel and producing your own charcoal. All of these subjects are covered giving detailed, practical advice based on my own experience, with a bit of theory when it seems necessary.

The designs of stoves and heating systems are comprehensively explained and these chapters provide plenty of basic, background information to help with the difficult task of deciding which stove to buy and whether it's new or second hand. Chimneys, flues and their associated issues are also carefully explained.

If you are a 'backyard technologist' then the 'home stove building' chapter could inspire you to look through the junk pile round the back of the workshop and get busy.

The way wood burns and how stoves are designed to take full advantage of the natural burning process is also explained and linked to the earth's carbon cycle.

Over the last six years I have had great fun introducing guests to the wood-burning cooker in the Ecolodge on our smallholding. They helped me understand just how many people have never had the opportunity to work and play with a wood stove or cooker and why using wood as a fuel is often seen as a bit scary and to be avoided. It is this lack of experience that keeps heating and cooking with wood as something of a smoke-and-mirrors 'black art', which, of course, it isn't. To try to overcome this I have included a 'beginner's guide to stove use' and 'how to cook with wood'. Once the basics are mastered it won't be long before you realise that stoves using properly seasoned firewood are a delight and behave with impeccable manners.

Although this is a practical guide to heating with wood it is written in the context of growing concern about global warming and an increased awareness of, and commitment to, reducing our own carbon footprint. Using wood as a fuel is an ideal way to begin this process; a carefully installed wood-burning heating system allows you to reduce bills for energy and contributes to a carbon-neutral existence. It is also a beautiful, cosy and renewable source of space and water heating.

background

I think it is important to view home heating and cooking with wood as an integration of domestic life with the wider environment. So this background section provides information about how we got to where we are and how wood burning, atmospheric carbon, and carbon footprint are linked together. This will explain the role of wood burning in the carbon cycle and the various stages of the burning process.

some social history

When I was a small boy, back in the 1950s, I remember the house being heated by an open, coal fire. Coal was, of course, the only form of heating for many houses and central heating was unheard of. The Second World War had virtually bankrupted the country. The standing timber stocks in forests and woodlands had been drastically reduced in the war effort and firewood supplies were minimal or non-existent.

The industrial revolution had left Britain with a legacy of coal smoke and filthy towns and we have all seen pictures of dour post-war towns and cities, with buildings blackened by the constant coal smoke and smog.

So it was totally unsurprising that the bright new future offered by natural gas from the North Sea during the 1960s was welcomed with open arms. The drudgery of hauling filthy coal into the house and lighting a fire every day was gone. People no longer needed to live with one room roasting and the rest of the house damp and chilly. The Clean Air Acts of 1956 and 1968 gave impetus to this new, cleaner, brighter, utopian future, where the dream was labour-saving devices for services at the touch of a button. People were striving to improve their standard of living and, along with motor cars, foreign holidays and fitted kitchens, came gas-powered central heating. The vision had a down side, in that open fires that required any form of labour were considered a sign of poverty. As a result of all of this 'progress' the general public in Britain became disassociated from the concept of using solid-fuel heating in the home.

During the early 1970s the first major cloud appeared on the energy horizon with the OPEC oil crisis. Since then the concept of unsustainable energy supplies, and rising energy costs, has been on the edge of public consciousness. Now, in the early twenty-first century, energy supplies are vulnerable and convenience and blatant consumerism have made a vast proportion of the population entirely dependent on cheap, readily-available, fossil-fuel based resources.

carbon

The growing concern and awareness about global warming means we are moving into an age where a reduction in the use of fossil fuels is increasingly important.

It is relatively simple to explain the warming process. Carbon-based, fossil fuels were laid down in a period of high temperature, moisture, and carbon dioxide – from the Carboniferous to the Permian Age There were swampy, warm conditions and plant life was prolific. These swampy conditions enabled huge quantities of carbon to be locked away in vegetation. This process was similar to that of peat formation, where anaerobic and acid conditions prevented decay. The carbon dioxide-rich atmosphere provided the basic building blocks for rapid production of plant material. The depth of this peat-like layer can only be imagined as it was finally compressed by geological activity into coal seams, sometimes two or three metres thick.

Climatic conditions slowly changed causing a reduction of atmospheric carbon and setting the scene for the conditions on Earth that we are accustomed to. Burning fossil fuels releases this locked-up carbon and results in increased atmospheric carbon dioxide.

Here's a simple experiment to illustrate how carbon is captured from the atmosphere. Take a clear plastic container with a lid that can be sealed. Put several layers of kitchen tissue in it and add enough water to make the tissue quite damp, then spread an even layer of cress seed on top. Take the lid and put a small hole in the top with a bradawl or something similar. The hole should be large enough to admit air but small enough to restrict the loss of water vapour. Replace the lid and seal the container (except for the hole), then weigh the container. Place the container with the lid in a warm, light place but not in direct, hot sunlight. When the cress has grown, weigh the container again. You will find the weight has increased. The only access to the container is through the small hole. The solution to this, and the answer to how plants grow, is that plants consist of air or, to be more precise, they take carbon from the air and use it to build plant material.

carbon footprint

Having your own wood burner can do little about carbon dioxide emissions on a global scale but burning wood for heating and cooking uses carbon that is active within the carbon cycle. In this way no new carbon is being added to that which already exists within the global carbon cycle. If the wood was cut and transported by non-fossil-fuel means then it could be considered almost carbon neutral. In this case the wood would have to be cut from a sustainable source (more about this later, see page 18). So, using wood for heat instead of coal, gas, or oil can reduce your own carbon footprint.

plant chemistry

In most instances plants grow by trapping the power within sunlight. This power is used, usually within the leaf, to split water into hydrogen and oxygen, and carbon dioxide into carbon and oxygen. The process is called photosynthesis and it is a fundamental part of the carbon cycle.

Here's a simple summary of the chemistry involved. Water consists of two hydrogen atoms and one oxygen atom; hence its chemical formula is H_2O. Carbon dioxide has one carbon atom and two oxygen atoms, and its chemical formula is CO_2. Basically, if you burn either hydrogen or carbon in sufficient air, which contains oxygen, you get water, or carbon dioxide.

Energy is released as heat when burning takes place, and to reverse the process energy is required to split water and carbon dioxide into their component atoms. This is what plants have evolved to use the power of the sun for.

An equally clever process plants have developed is to recombine these atoms to produce a carbon/hydrogen chain and oxygen. The carbon/hydrogen chain is commonly referred to as a hydrocarbon and is a basic sugar that can be transported around the plant. The sugar is used for plant maintenance and the building of new plant material. In trees this new material is in the form of cellulose and lignin, along with other things. The cellulose is the main constituent of the cell walls and the lignin is the glue that binds things together.

So, in essence, the plant turns carbon dioxide and water into sugar and oxygen. The sugar is then used to make the plant material, called cellulose, which, in trees, produces wood. The oxygen produced in photosynthesis is released into the atmosphere where it is used in respiration and burning to produce carbon dioxide (see fig.1).

You can see, then, that this is part of the carbon cycle and shows how carbon is constantly being recycled. The diagram below is simplified and the hydrocarbon does not accurately show the structure of cellulose, but it gives you the general idea. There will be more information about cellulose later when we cover drying timber (see page 37).

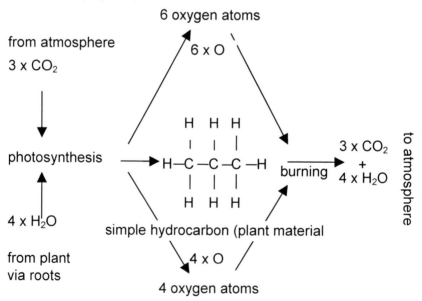

fig. 1: carbon/oxygen cycle for growing and burning wood

how wood burns

As we all know, to light a fire you start with small size material, which includes paper and kindling; lighting the paper heats up the kindling so that it starts to release some of its hydrocarbon gases. These gases then ignite and keep the process going. There are actually three phases in the process of burning wood.

phase 1
The heat from wood that is already burning heats up any water within the igniting log. This water then evaporates and is lost up the chimney. As it evaporates it takes with it a considerable amount of heat. Basically, it takes plenty of heat to raise the temperature of things, especially when you want to change their state from liquid water to steam. The evaporation of the water keeps the temperature down. This effect can be demonstrated by applying surgical spirit to the skin: as the spirit evaporates the skin feels cold.

phase 2
Once the moisture has been driven off the temperature rises and hydrocarbon gases are released. These can burn if there is sufficient heat and oxygen. Half of the heat potential of wood is in these gases, so efficient wood burning depends on the delicate balance of airflow through the burning logs.

phase 3
The third phase is when all the gases have been removed from the log to leave charcoal. This charcoal then burns with no visible flame but it has a voracious appetite for oxygen.

From these three phases, which happen simultaneously, we can extrapolate various basic rules for efficient wood burning.

phase 1
The wood must be as dry as possible. The wetter the wood the more heat is used to evaporate its water content and, therefore, lost up the chimney. Any water vapour produced will keep the temperature down, which causes the wood gases to condense and create tar in the chimney. This is an alert to a series of large and potentially bad things, (see page 75). Dry wood enables the temperature to rise rapidly and so gases are produced freely.

phase 2

For the gases to burn freely there should be air available just above the logs. Ideally it should mix with the gases and have enough heat for the mixture to ignite. Many books state that a temperature of 540°C is needed for this process to happen, which means very little to most of us. What you need to know is that the charred logs need to be glowing on the outside. When the air vents of a stove are open on a lit stove there is a rush of air through the vents. This causes turbulence within the stove and so mixes the gases and air together. Stoves are designed to create this turbulence in the hottest part of the fire and then almost compress the gases as they make their way to the chimney. Baffles are used to direct and reduce the velocity of this gas mixture to produce the most efficient burn.

phase 3

The charcoal has the potential to give half of the possible heat output of the wood. For it to release all of this heat there should be plenty of air available to complete the burning process. This means that the stove should be designed to have a separate air supply to the charcoal layer, so that it does not impinge on the air supply to the burning gases. It will also affect how the operator controls the stove to give efficient and clean burning.

sustainable timber production

We are all increasingly aware of the impact the human species has on the rest of the environment. It is in our own interest for our supply of food, goods and energy to be sustainable, otherwise we will use up all the resources available and then die out as a species and take many other species with us. Some people think we are already well along this path. So, if the production of timber is not sustainable then there will be none left for future generations. There will also be a lack of habitat for many of the species on which we unknowingly rely to keep the biome healthy. You cannot rely on agribusiness to supply a rich, diverse and chemical-free environment. The essence of sustainability in forestry is to cut less timber than grows each year and always replant after felling, except where the coppicing system is used. There is more about sustainable firewood production in the chapter on 'growing your own fuel' (see page 45).

In Britain at the moment we have the Forestry Commission, see 'resources' (page 141). This organisation has the responsibility for governing the felling of trees in woodlands, through the felling licence system. The felling licences are issued on the condition that woodland areas are restocked (replanted). It is in the interests of forest managers to restock felled areas because it is in the nature of forestry to take the long view. It is their job to provide timber for generations to come and, besides, there are grants available for replanting.

The action of arboricultural companies can lead to the loss of tree cover in urban areas, but there is little control over this type of felling unless a tree preservation order (TPO) is put on the trees. These orders are controlled by the tree officer at the county council and cover not only the tree but also the ground it stands on. This means if a tree dies then it is a, legally enforceable, duty of the land owner to replant, but only if someone cares enough to enforce it. There is another type of blanket protection, called a conservation area, which refers to conservation of villages in respect of trees and buildings.

Felling licences from the Forestry Commission are not required for gardens. If you are looking for sustainable firewood then that supplied by tree surgeons cannot be classified as sustainable because, in the majority of cases no replanting takes place.

buying and storing wood

So how do you buy firewood? Well, the answer is that there are many different ways possible and the way you choose will depend on a variety of things like price, the amount of storage space you have, what is available in your area, how much effort you want to put into preparing your own wood and so on. What you need to have available to use is well-seasoned wood with a moisture content of twenty per cent or less (for more about moisture content see page 38). That's it really; the rest is about price, volume and convenience.

I am not giving any guide prices in this chapter as they will be out of date in a year or so. What you have to recognise is that the more processed the wood you buy is, the more expensive it will be. A firewood merchant will probably buy timber in by the lorry load, by weight, and sell it by the truck load, by volume. So not only does he have to pay up front for the timber but he also has to allow for conversion to firewood and delivery. This conversion and delivery takes considerable effort and expenditure in equipment.

It is also very difficult, when you first start, to predict how much wood your system will use annually as there are so many variables, including the size of your house, the quality of the insulation, whether you also cook with wood, how many stoves you have, how cold the winter is, and so on. Our woodshed holds about thirty-six cubic metres of stacked logs and that lasts us most of the winter, for a three-bedroom cottage in an exposed situation. The only further thing I can say is, if your woodshed is big and you don't use all the contents in one year then that's a bonus and there's more left for next year. It won't take long for you to work out how much firewood you need in stock and the more stock you have the longer it has to season and the more heat you will get out of it.

different methods of buying wood

Probably the least cash-efficient method to use is to buy a bag of logs from the petrol station down the road. I have seen some bags of awful logs sold in garages and, of course, they would not last an evening. A local firewood merchant may supply these logs but you pay extra for bagging plus the mark-up for the petrol station.

A more cost-effective way to buy is to get a 'load' of wood from a local firewood merchant, see 'resources' (page 141). The merchants that have been in business for some time have an interest in keeping customers, and should give consistent loads. If you want a load of logs for immediate use then the moisture content will need to be low so it is best not to buy by weight as this wood, of course, is lighter than fresh-cut firewood and will cost you more. It is fair enough to pay more for seasoned firewood but I would recommend using a moisture meter to check what you are getting. Asking the merchant for seasoned firewood can give variable results – from honest answers to bending the truth beyond all recognition. For example, you could ask how long since the wood was cut and split and whether it has been kept in a dry place, but bear in mind that the answer could refer indirectly to the felling time rather than when the wood was split, giving a spurious answer whilst not actually lying.

So, if you don't buy by weight, how should you buy? The best way is by volume of stacked logs. You can work out the volume of a level truck load of logs, and work it back to a price per cubic foot or cubic metre of log load. It's really the only way to compare different merchants and different truck loads. The species of tree cut into logs also matters: more about species and heat output later (see page 43).

To do these calculations it is useful to know what the stack volume of each merchant's truck load is and you can then use the stack measure to estimate the volumes of logs as follows. When the timber is fresh cut and green the approximate volume to weight ratio is 2 to 1. This means, for example, that a stack of 2 metre logs that is 4 metres long and 1.8 metres high has a volume of $2 \times 4 \times 1.8 = 14.4$ cubic metres of stack, and gives an approximate weight of 7.2 tonnes (14.4 divided by 2).

The same process can be used for cut firewood although there will have to be some approximation and 'rounding up' of the volume if the stack is not level across the truck body.

I know a merchant who has solved this problem by using large vegetable bins that hold about 1.6 cubic metres of logs. He has many of these bins that are filled and stacked, up to six high, throughout the spring and summer. The timber dries well and the customer is guaranteed a standard volume of logs for each delivery.

An even more cost-effective method to use is 'forward buying'. This usually involves buying all of next year's firewood during the current year, with delivery out of the main supply season, and will only work if you have enough space to store and dry the timber and enough money to pay for a year's supply.

You buy green (fresh-cut) firewood and will be able to negotiate a reduction in price for unseasoned product, taking delivery in the summer and buying a year's supply in one order. The benefits are considerable on both sides. For you there is the reduction in price and plenty of time for the firewood to season (dry out) to twenty per cent moisture content, or less. The money is well spent because of the discount and there is no disputing the fact that dry timber gives shed loads of heat compared to unseasoned timber. It is also beneficial for the merchant as they have income during a lean time and can deliver in a more relaxed manner: the days are longer, the weather better and they don't have the pressure of orders all coming at once. It is also accepted that the timber is green so there are no disputes about moisture content.

Another method you can use is buying firewood 'uncut' on an industrial scale. You buy by the lorry load and the size of the load depends on which area of the country you live in. It also depends on the size of the lorries available, cost of road fuel and the individual haulage contractors in the area. In Lincolnshire, for example, the best way to buy uncut firewood, also known as round wood, is by the articulated lorry load, which yields anywhere from 20 to 26 tons. I know I have just said 'don't buy by weight', but the rules change once you are dealing with industrial volumes. Weight is the only way for this type of load to be quantified. The wood is usually freshly felled – it is almost impossible to buy dry timber by this method. The reason for this is as follows: it is in the interest of forestry contractors to get the timber sold rapidly as they are paid on weight tickets supplied by

fig. 2: forestry contractor hard at work

the haulage contractors and won't get paid until the timber is moved. There is also another problem; if the timber is kept for any significant time, it loses weight through drying out. This represents a reduction in payment to both forestry and haulage contractors. In the right conditions timber can lose forty per cent of its weight in six months. Most lorries have on-board weighing systems so it is easy to measure the size of the load, otherwise they have to be weighed on a public weighbridge. The haulage contractors do this and provide a printed weight ticket for the load.

If you buy firewood by this method you will need space to stack the timber. As a rule of thumb a stack would be two metres high, two-and-a-half metres deep, and nine metres long. Buying your timber this way gives you several years' supply but it is green so you will need firewood in stock to tide you over for the first year.

If keeping the cost down is a priority and your time is restricted, it is always worth looking in the phone book to find local sawmills – some years ago we bought all our firewood from a local sawmill. This mill dealt mainly with hardwoods, cutting anything from mining timber through to joinery quality. The 'slab wood' waste was strapped into one-metre-diameter bundles, the bundles were

fig. 3: timber lorry unloading firewood

about two and a half metres long, and then sold off. I can almost hear some of you saying 'What's this slab wood you're talking about?'. Well, imagine you want to cut a gate post out of a log, the action of cutting a round log into a square gives some waste. This waste has a curved side with bark attached and a sawn side and that's slab wood (see fig. 4). The size of the slab depends on the logs and the product they are cut into. Hardwood logs vary in size and produce more waste when cut which results in larger slab wood pieces than with softwoods. The bundles we bought always had a variety of content, size of material and species. The buyer could either collect or have several bundles delivered. We used to buy about six bundles a year, which provided most of our firewood and also provided odd pieces which could be used around the smallholding.

If you live in a town and access for lorries delivering wood is difficult, for instance if you live in a terraced house with access only through the front of the house or limited access to the back, it would be best to install a stove that uses pellets rather than logs as the pellets are delivered in bags and are easier to handle and store. If you have limited access and need the delivery driver to wheelbarrow logs from the lorry you should discuss this with the supplier when ordering as there may be an extra charge.

Sourcing a reliable supply of firewood is all about getting to know local sawmills, timber hauliers, and firewood merchants, see 'resources' (page 141). This can be achieved through trade associations like the Forestry Contractors Association, the Royal Forestry Society, see 'resources' (page 141), or indeed through local newspaper adverts. If you are interested in woodlands and timber, (including firewood), the Royal Forestry Society has regional groups and they organise visits to many woodlands that are not usually open to the public. Valuable contacts can be made through these visits and their quarterly journal provides all-round interest, see 'resources' (page 141) for details.

fig. 4: slab wood in the firewood stacks

cutting your own wood

The ways of obtaining firewood are various and wondrous. Those described above are the common and easy-to-set-up methods.
You can, of course cut your own, but this is a very different approach and involves lots of labour.

Felling trees is a potentially dangerous activity and should not be undertaken lightly. The use of chainsaws is also very dangerous and chainsaws should not be used without proper training and correct PPE (personal protective equipment). As you can see, felling trees with a chainsaw puts two dangerous things together and, as we all know, two dangerous things don't make one safe thing.

chainsaw training

I'm not about to recommend that anyone should use a chainsaw, because of the modern disease of 'litigation culture', and, just to make sure the message is clear, I must emphasise that you shouldn't use a chainsaw unless you have received recognised, quality training.

In Britain the governing body overseeing chainsaw training is Lantra, see 'resources' (page 141) for details. Their courses are registered with the Qualification and Curriculum Authority (QCA), see 'resources' (page 141), and so are set at NVQ Level 2. The course content is also approved by the Health and Safety Executive (HSE), see 'resources' (page 141). For anyone who wants to use a chainsaw in a safe and efficient manner, then it would be of significant benefit for them to book a place on a course. On the basic, two-day course you receive training in safe procedures, machine maintenance and chain sharpening, and cross cutting. Other courses are more in-depth and include felling trees and taking down trees that get hung up on other trees as they fall.

I can't emphasise the importance of training enough. I have been involved in training many people to use chainsaws for years and they get huge benefit from the underpinning knowledge provided, about maintenance and sharpening. The difference between a sharp chain and a slightly blunt chain is enormous. Not only is a sharp chain a lot less effort to use, it causes less wear and tear on the machine, uses less fuel and oil, and is a lot safer. One thing that many people do not realise, for instance, is that if the chain touches the ground when cutting this immediately blunts the chain: the movement drags soil into the cut and this then acts like sand paper and removes the cutting edge. The Lantra website, see 'resources' (page 141), provides a register of qualified chainsaw instructors.

Warning! If you are a mechanical or practical half-wit then stay well clear of chainsaws.

storing firewood

There is no point having top-notch firewood if you leave it outside in a heap to get wet. I can't reiterate this enough: get the wood dry and then keep it dry. This is the ideal and, as with all prescriptions for best practice, should be a goal to aspire to. Your methods will depend on you, your infrastructure and the amount of effort you are prepared to put in.

The firewood we burn is left in the stack in log lengths for at least a year, but some of it has been there for two years. It is then cross-cut and stacked randomly in the firewood shed. This shed has three bays, with an internal volume of about twelve cubic metres per bay. As a bay becomes empty I start filling it again, so there is a constant rotation with plenty of dry timber in stock.

What happens in practice is that at the beginning of the autumn all the bays are full. Sometime in early December the first bay will be empty and I will start to refill it. It's a steady job that doesn't need rushing because these logs may not get used until the following September. Having said 'no need to rush', there is also no advantage in leaving it until later in the year. The object of the exercise is to get partially air-dried firewood under cover to dry during the following spring and summer.

The shed has to have some special attributes: the roof should, of course, be waterproof with rainwater taken away by gutter and land drains to prevent soaking the ground.

The sides should be permeable, to allow the wind to whistle through. Air movement through the stack is vitally important to clear moist air out and replace it with drier air that can absorb moisture from the stack. This is why the firewood is stacked randomly so that there can be good air circulation – and also it's a pain to stack it ever so neatly. The shed sides could be vertical boards with a gap between each board. This is up to you and the materials you have to hand: hazel or willow hurdles would also do very well. There also needs to be plenty of air circulation at ground level and building the floor of old pallets is a good way to achieve this (see fig. 5).

Using this method means we dry our firewood to the best of our ability and we always have plenty in stock. There is nothing worse than a log beginning to sizzle just after you have put it in the stove; it means wasted heat and tar production.

In the 1940s there was a young engineer and pilot called Frank Whittle who was concerned about the efficiency of propeller-driven aeroplanes, which were quite inefficient and couldn't fly fast or high. He ended up designing the jet engine, but the point of my story is that all he could hear when listening to aero-prop engines was the sound of wasted fuel and energy. It's much the same with sizzling logs, except you don't have to invent an entirely new technology to overcome the problem.

fig. 5: wood sheds with plenty of ventilation for drying.

Here's an example of not getting it quite right: I know someone who cut some firewood from logs that had been stacked for some time – so far so good. These were then stored in a small brick-built shed – I think it was the old privy. When they came to use the firewood the following winter it still contained plenty of moisture and did not give the heat that was expected or required.

The story starts off right and they had all the right intentions, but the shed didn't allow the firewood to keep drying because of the restricted airflow.

more about wood burning

To get high-moisture-content firewood to burn, you have to open up the stove air vents, which encourages the fire to roar away. If you shut the stove down the wood just sits there and smoulders, or goes out. Opening up the vents sends lots of the heat up the chimney, which is then lost to the atmosphere. The net result is burning a great deal of fuel and a cold house. Dry wood, on the other hand, burns gently in a restricted airflow giving huge quantities of heat, and reducing the flow of warm air into the stove from the house. The hydrocarbon gases are burnt off because there is ample turbulence in the firebox, without all the heat rushing up the chimney.

The settings of the stove controls should be:
Air vents partially open and chimney damper partially closed to prevent all the heat being sucked up the chimney.

I think I have hammered the message home.

firewood processing

Let us assume that most people will do a cost-benefit analysis and decide to buy their firewood ready cut. If they are sensible then they will buy in advance and so always have plenty in stock.

If you, however, are going to cut your own then there are plenty of methods you could use, which include:

- hand saw
- reciprocating saw
- chainsaw
- saw bench
- firewood processor
- splitting

hand saw

There are a couple of types of saw to choose from, either a traditional-type carpenter's hand saw, which is fine if you look after and know how to sharpen it, or a bushman's bow saw with crown-type teeth. The blades on these are almost always induction hardened and so you cannot sharpen them but replacement blades are cheap and last a long time, if they don't get rusty. The rust bites into the cutting edge quickly and so all saw blades should be oiled on a regular basis before the saw is put away.

reciprocating saw

I've seen several of these, which used to be made commercially. The basic idea is a large saw blade, which is mechanically driven in a sawing action (see fig. 6). Imagine if you can an engineer's mechanical hacksaw, same sort of thing. Load a log and set it going. The idea is fine if you have large logs, or are pottering about and are at hand to re-set at the end of the cut. You could

fig. 6: reciprocating saw

always build one to take several logs at once, perhaps held in place with a ratchet strap. The benefits of this method are that you could use a relatively low-powered electric motor, about three-quarter horsepower, and that it's a lot quieter than a saw bench or chainsaw. Isn't it a pain when someone starts up a noisy two-stroke engine on a peaceful sunny afternoon?

chainsaw

We've talked about these before (see page 27); the same warnings about safety and training apply and you should also bear in mind the noise they make. There is no point alienating yourself from the rest of the community by being anti-social. Electric chainsaws are relatively low-powered and a lot quieter but the cheap ones always have problems with the chain-oil mechanism.

saw bench

As with all powered tools these are unforgiving to the novice and should be used with care. The major problem with circular saws is that they can grab at a log with a violent action. There are several saw benches suitable for cross cutting logs. Some have a swinging trough that holds the log, which enables the operator to introduce it to a fixed, circular saw blade. The trough reduces the grabbing effect by holding the log more firmly than can be done by hand.

The saw bench I use is an old McConnell-tractor-driven unit (see fig. 7). The unit is driven from the tractor power take-off, and has a sliding table. This is ideal for cutting up big rounds of timber or those horrible knotty bits, which are impossible to split. The sliding table means you don't have to slide the logs over the table, the whole lot moves relative to the blade. This way there is less chance of the blade grabbing the logs. There is a tendency for the log to start rolling as the blade starts to cut, so you have to have a good grip on the log. On this model a pivoted guard comes over the top of the log with timber dogs on the underside to grip the timber. There is also a guard to cover the blade. I must admit it's a bit of a beast with a blade seventy-five centimetres in diameter. The general rule is: keep the guards in place and keep your fingers out of the way.

fig. 7: tractor driven saw bench (guard removed for clarity)

firewood processor

These are used by firewood merchants and reduce much of the double handling of material. The logs are introduced down a conveyor to the cutting area. The logs are cut either by a pivoting chainsaw bar and chain, or by a circular saw that moves across the heavily-guarded cutting tunnel. The off-cut then drops down to where a hydraulic press splits the log. The processed firewood then makes its way up an elevator and into bins or the back of a truck. These machines are expensive and are usually driven by a tractor. I include them in this description for extra background information, and also because it may be worth estates or collectives considering buying one given the annual volume of firewood they need to cut.

splitting

Once you have cross-cut the firewood, it may be that some of it will be still too large. The rounds, otherwise known as rings or chogs, then have to be broken down to a size that will fit into your stove. With large rounds I use the saw bench, mainly because my shoulders are about worn out, and will not deal with the shock of using a splitting axe. For fitter people a splitting axe is a useful tool. The difference between a splitting and felling axe is the profile. The felling axe has a narrow profile designed to cut into the timber and, if used for splitting logs, tends to get stuck in the end grain (the cut surface). The splitting axe is a cross between a sledgehammer and a wedge. The extra weight of the large head keeps the momentum going to help encourage the split to develop. The profile of a blunt taper prevents the axe getting stuck. It either creates a split, or bounces back a few inches.

fig. 8: axes - splitting axe (left), felling axe (centre), hatchet for kindling (right)

The size timber is split down to will vary according to the intended final use of the wood and the size of your stove: you need large blocks for slow burns, smaller size for boiling the kettle or getting the heat built up in the morning.

Another manual way of splitting particularly nasty rounds is to use steel wedges and a sledgehammer. A wedge can be driven into the timber to create a split, which is then followed by another wedge to increase the split. Some timbers are particularly difficult to split, but most problems are encountered with knotty pieces.

In case you don't know, a knot is formed in a tree where a branch grows. The knot creates interlocking grain patterns, which resist splitting. Splitting the wood gets more difficult as the timber dries, so knotty bits need to be split by hand when green, or a mechanical method, like my saw bench, should be used.

A splitting axe is used in conjunction with a chopping block (see fig. 9). I find it is best to have the block slightly higher than a wheelbarrow, so that, with careful positioning, the split blocks can sometimes fall directly into the barrow. The other advantage is that it reduces the constant bending down to reposition logs.

kindling

In the last section I mentioned size of firewood for different types of heat and an extension of that thought leads us to kindling. This consists of small, split pieces, about the thickness of a finger, which is used to light a fire or rekindle from glowing embers. Our firewood stack contains many different species of timber and so I tend to sort out the best bits for kindling and put them to one side. By 'best bits' I mean the pieces that are relatively knot free and have straight grain. The softwoods make good kindling, as do species like Leyland Cyprus, Willow and Alder. The main criteria are that the timber will split easily and is dry before use. We have a large, galvanised bucket that is kept full of kindling, so there is never an issue about having to split kindling first thing in the morning – which is a pain.

fig. 9: chopping block at a suitable height for a wheel barrow.

how timber dries

wood structure

Wood is constructed from cells that are primarily made up of cellulose, a compound of hydrogen, carbon and oxygen. The majority of these cells form upright tubes – which is a bit of a generalisation, but will suffice for our purposes.

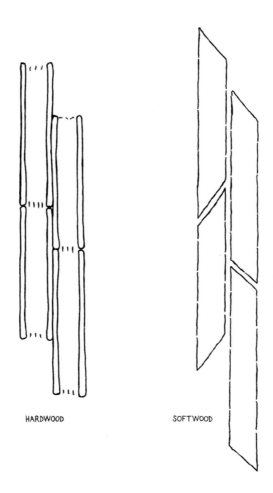

HARDWOOD   SOFTWOOD

fig. 10: hardwood and softwood structure

In hardwoods the tubes are stacked end on end and go all the way up the tree. In softwoods each tube is in the form of an extended lozenge (see fig. 10). The purpose of these tubes is to distribute water and nutrients throughout the whole tree. These tubes have little pores in their side walls to allow the passage of liquid sideways into a neighbouring tube (see fig. 11). As you can see from fig. 10, the hardwood cells have perforated ends called sieve plates, through which the liquid flows. In the softwood cells the only flow is through the pores in the cell walls.

When the living wood is freshly felled these cells are full of water and the water can be divided into two main types:
Free water: this is the water that is lost when firewood is seasoned. It is found within the cell enclosure.
Bound water: is water which is found within the cellulose bonds in the cell walls. It is the loss of this water that makes timber shrink and is the bane of all carpenters' and joiners' lives.

moisture content

The moisture content of wood sometimes confuses people and always causes discussion on courses. There are several ways of determining moisture content, but the common way is by using the 'oven dry' basis. The weight of the wood is taken before it is dried gently in an oven, so explaining the term 'oven dry' basis. The oven-dried weight is then subtracted from the original weight, to give the weight of the water lost. The weight of water is then divided by the weight of oven-dried wood, and the answer is multiplied by one hundred to give the percentage. Should the weight of the water and wood be equal then the answer is a hundred per cent. This explains why you get some timbers with a hundred per cent moisture content when they are freshly felled, for example Norway spruce or Poplar. I ran a staircase company some years ago where we kiln dried our own locally-grown and sawn timber. One of the favourite timbers was Poplar because it grows to large sizes and is inexpensive. When unloading the freshly sawn timber we could only carry one board each, but once it was kiln-dried we could carry three. That just shows that the moisture content was at least a hundred per cent.

Many of the moisture meters on the market are not very accurate at higher percentages, but this doesn't matter for our purposes because content only really matters in the drier stages. So don't get hung up on the numbers, just dry the timber properly to get a quality heat source.

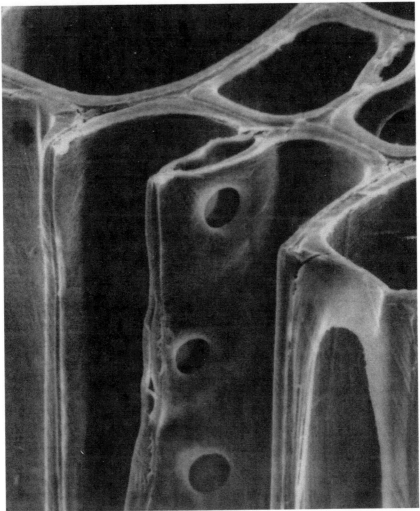

fig. 11: cell structure

cell structure and drying

The cell structure of timber affects the way wood dries. The drawing showing cell structure (see fig.10) shows pores in the cell walls. These pores occur in both hardwood and softwoods, but the softwoods have a valve in the pore that closes if air gets into the cell on one side. The valve is called a torus and is held in place by fine filaments. This is a defence mechanism to prevent fungus and bacteria from entering the tree when it is damaged. It has the extra effect of reducing water loss and increasing drying times in cut timber.

Different timbers have distinct properties that affect drying:

ring porous timber
This will only allow movement of water along the annual rings, which means the water can only escape from the ends of the annual rings on the cut surfaces. Oak, Sweet chestnut and Elm are some examples.

diffuse porous timber
This will dry both across and along the annual rings, which allows the timber to dry much more quickly and evenly. Willow and Poplar are some examples.

semi-diffuse porous timber
This is ring porous with a small amount of porosity across the annual rings, it dries quicker than ring porous timber. Birch and Cherry are some examples.

To dry firewood we have to encourage the moisture from the inside of the log to migrate through the outer cells, and from there to the atmosphere. When kiln drying timber for commercial joinery we were always aware of 'case hardening'. This occurs when the outer cells become so dry that moisture is prevented from migrating from the inner cells, and is caused by drying the timber too quickly. This won't occur with air-dried firewood, but illustrates the natural process involved.

In practice when you are using ring porous timber it means the drying times are not only longer, but to achieve effective drying in the middle it is best to split all of the wood, even if its small, branch wood.

bark

The bark slows down the drying process, but I am not suggesting that the bark should be removed as splitting the firewood enables quicker drying. Ash, Cherry, and Birch are particularly problematic: Ash poles, when young, hold on to their smooth bark and this prevents drying. This is a symptom of the wood's relatively low moisture content, which means that when Ash dries there is little shrinkage in the timber. The knock-on effect of this is that the bark does not crack and so it holds any moisture in. Birch bark is waterproof and used to be used for canoe skins. The bark holds in the moisture and encourages rot to start. Birch is good firewood but must be spilt to release moisture, and when it is burnt gives off a peat-smoke smell – redolent of the Highlands.

A friend of mine told me about a company down in Kent that used to buy Birch poles to make broom heads. The company insisted that the birch poles had three strips of bark removed along the whole length. This was to ensure that the pole dried out before machining. When I put four-foot Birch logs in the outside stack, I always run the chainsaw through the bark down their length to give the same effect.

annual ring structure

Trees grow at differing rates throughout the year, which produces the annual ring structure in the timber. There is also a difference between hardwood, mainly deciduous trees, and softwood, mainly conifers. Most softwoods keep their needles through the winter and so continue to photosynthesise at a slow rate. This means that they keep growing, even though it is at a slow rate. Hardwoods lose their leaves and so go into non-growth dormancy.

The annual ring structure is as follows (see fig. 12): The lighter coloured ring is called the spring wood, which, of course, is produced in the spring. The trees leap into life after the winter dormant season and grow quickly. This produces large cells with relatively thin cell walls, hence the lighter colour.

fig. 12: ash that has grown quickly showing annual rings

The darker rings are the summer wood, where growth is reduced in favour of reproduction. The canopy of leaves is in full production, and all of this energy is going into producing flowers and seed.

Trees that grow slowly produce dense timber with smaller cells. This denser structure has a greater resistance to the movement of moisture across the rings, particularly through the summer wood. This is why ring porous wood is found in the denser timbers like Oak and Ash. The moisture can only move through the spring wood, and so can only move around the annual ring.

species

I have mentioned different species and some of their properties and although there are many books which give detailed lists of their specific uses, the most important thing to remember for our purposes, beyond moisture content, is that, within reasonable bounds, all timbers have about the same heating value by weight. Woods are made of the same stuff, it's just some are denser than others. So the dense ones burn more slowly and the weight is contained in a smaller volume. The timber you choose to burn for a specific type of heat is dependent on the density, assuming the moisture content has been effectively reduced and the firewood has been dried for sufficient time.

growing your own fuel

Growing and cutting your own wood fuel needs a medium to long-term view. It will be some time before you see any fruits of your labour in heat form. There are many other benefits to be gained as well as harvesting your own fuel as I will explain as I go along.

The ideal situation is to plant a new wood and the reasons for doing so are many fold. I realise this is a best-case scenario but it gives us all something to aspire to.

If you, or a group of like-minded people, plant a new wood on agricultural ground then the first thing you are doing is wresting a bit of the countryside back from the 'agri-desert'. The size of the field, of course, depends on cash resources, how many members in the group, if any, and what land is available locally.

The wood you plant will be taking up new carbon and so it will be a truly carbon-neutral fuel source. Let's look at that statement in a bit more detail. An existing wood contains a fairly stable volume of carbon over the medium term. If the wood from existing woodland is used for fuel you are returning carbon back to the atmosphere that is already in the carbon cycle. If the fuel comes from new woodland, however, then this new wood has theoretically taken up recently-emitted, locked-up carbon that was not in the carbon cycle, fossil fuel derived. In this way you can have a direct effect on your carbon footprint, beyond reducing consumption, and it goes a long way to offsetting the carbon emitted in everyday, modern life.

As it grows the new woodland provides a huge habitat benefit for wildlife and a pleasant leisure and kids' learning resource. There is an important ecological aspect to habitat within the landscape, which is called contiguity. This is to do with wildlife corridors; contiguous refers to 'joined together', where hedgerows, woodlands, grass fields and ditches provide a joined-up pathway for wildlife to move along. We are not only talking about animals, insects and invertebrates, as plants also move surprisingly quickly sometimes. Now, I am not suggesting that a Silver birch will creep up on you if you stand still for long enough, but birch is a pioneer species and will self-seed remarkably well given an area of reduced disturbance. Planting a new wood on agricultural ground is exactly that, it reduces the level of ecological disturbance.

grants

The information in this section will be brief because it will no doubt be out of date in the blink of a politician's eye.

The agricultural grants have changed from a production base to an ecological base and so grants are, currently, available for habitat benefits, agricultural reversion (taking land out of agricultural use), grassland and for land which has a 'low nutrient status' – so that the land is used but is not fertilised in any unnatural way.

Forestry grants are available for new planting and are awarded on a points-scoring system, which includes size, contribution to local biodiversity, species, habitat action plans. Local wildlife trust groups should be able to help you with these.

You will need to find out what's available at the time you are planning your wood. I am just pointing out that there may be grants available.

planting

The planting season starts with the dormant season, which is usually early November, depending on global warming. In forestry we plant bare root 45-60cm-sized trees, which are small and establish more easily than larger trees. They are also relatively inexpensive, depending on where you buy them: it is best to get them from a dedicated tree nursery. In lowland Britain the use of tree guards is essential and so the cost of planting can be divided equally into three: tree, guard and stake. The guards are essential to prevent rabbit, hare and vole damage. Hares nip the trees off, rabbits eat the trees or remove the bark of larger saplings and voles remove the bark of newly-planted trees. All of these forms of attack can kill the tree or reduce its vigour and ability to establish.

When planting trees on any area larger than, say, a quarter of an acre, it is best to plant them so that you will know where they are when mowing or weeding, even before they are big enough to be seen above the tall grasses and weeds that will inevitably grow. It is a good idea to plant them in gently curving and meandering rows so that all the rows are parallel and follow the same meandering line. This means that the lines are not obvious from a distance, but are evident close up for mowing or weeding.

The new trees come in bundles of twenty-five and are packed in bags. It is essential that the bags are stored in a cool place away from frost or direct sunlight. This keeps the trees in best condition until planting, which should be as soon as possible after delivery or collection. When planting keep the trees in a bag out of the sun and wind until the hole is dug or the planting notch is cut. The roots are sensitive to drying out, so don't wander about with bare root trees wafting about in the breeze.

weeding

It is essential for good tree establishment to reduce the competition from surrounding vegetation. The forestry industry best practice is to keep a weed-free area of about one metre around each tree. How you do this is up to you and the time you have available. Weeds and grasses grow rapidly and can shade out small trees in a matter of weeks. Some of the methods to choose from are organic mulching, mulch mats, (like old carpet or cardboard), or a systemic chemical control that kills the whole plant and not just the top and doesn't stay in the soil. Roundup is a product that is widely used in forestry. If you do this it is absolutely essential to keep the control up for at least three years. The difference between weeded and un-weeded planting can be astounding: from no growth at all to trees that are three times as high.

species choice

There are certain tree species that have a growing strategy ideally suited to growing in open areas. These are called 'pioneer species', and they grow rapidly in the first few years and then slow down to steady growth. Birch, Ash and Sycamore are good pioneer species. Other species grow more slowly to start with and then speed up in later years. When planting a wood specifically for fuel a mixture of species is important to give differing densities of firewood and, hence, differing types of heat. I would suggest a combination of Ash, Birch, Alder, Willow, Beech, and Oak.

Each species also grows best under certain ideal conditions known as silvicultural characteristics. For instance Willow and Alder do well in damp conditions so you would not plant them in a dry, chalky soil, but Beech would do quite well in such situations. A bit of research is required before you plant, and it is also a good idea to have a look in the local area and see what is already growing well. That will give you a good idea of the possible range of species to choose from. The quick-growing trees initially change the microclimate from open field to low woodland. Trees grow better in a woodland environment, so this change then benefits the slower-growing species.

harvesting

The traditional way of harvesting firewood is called coppicing, which involves regularly cutting the trees back to ground level to encourage the growth of side shoots for future use.

The ideal way of working is to divide the area of woodland into a number of sections that is equal to the number of years of the rotation. For example, let's say we are going to cut each section every ten years and the wood is 5 acres, this means the yearly area to be cut is half an acre. When cutting for firewood it is best not to let the diameter of the trees get too large, or there will be a lot of splitting to do. If you are trained to use chainsaws, and want to, they are useful for coppicing, but it is entirely possible to use a bow saw on trees up to twenty centimetres in diameter. It depends on how much time you have.

Once you have felled an area then the next year the stumps of the felled trees will begin to send out shoots and re-grow. This time there will be a multitude of new shoots and so the trees become multi-stemmed. The trees grow faster as the roots are now well established. This is how short-rotation coppice willow is produced for fuelling biomass power stations: the rotation used is between two and three years to give diameters of up to two inches.

Willow is established by pushing cuttings into the ground, but it is grown as a monoculture for fuelling power stations. It is possible to add some Willow to your planting mix to get some quick-production growing, and to provide some side shelter for the other trees, but it must be carefully managed or the willow will end up shading out surrounding trees.

managing existing woodland

If you are harvesting your firewood from existing woodland you should start by surveying what species, age, condition, and form already exist. You are trying to identify the structure of the woodland and what products are growing.

The variety of products depends to a large extent on the vibrancy of the local markets and economy. Major products are firewood and sawlogs, with hazel binders for hedge-laying and bean sticks as extras. Sawlogs are exactly what the name suggests in that it is timber suitable for conversion in a sawmill. There is a great deal more that could be said about this and I cover it in detail on LILI's Woodland Management course see 'resources' (page 141).

Let us imagine a typical under-managed, lowland woodland that has been left to degrade in favour of shooting. There will be some trees that are tall and relatively straight and could be suitable now, or later, for sawlogs. Between and under this patchy canopy there will be a collection of other trees that are usually of poor form and the crowns are suppressed or deformed through trying to grow to the light. These lower trees, and any other canopy trees that are damaged or of poor form, should be used primarily for firewood.

One method of producing a regular supply of fuel is to coppice a regular area every year see, 'harvesting' (page 48). When there are canopy trees within the annual felling area you will need to make a choice to either fell these trees or leave them until a further rotation. There is, however, an issue with light levels, in that the cut stumps of the trees need good light to produce strong, healthy, new growth. The size of the felling area relative to the to the number and crown size of retained trees becomes important here. If you are primarily interested in producing firewood then err on the side of increasing the light levels on the woodland floor and remove the canopy trees. Opening up the woodland floor to light also has great habitat value for plants, insects and birds, and mimics the action of natural disturbance (disturbance ecology) where the action of succession returns a damaged area of woodland back to high forest. For further reading see 'resources' (page 141).

charcoal

Many households use charcoal for cooking outside in the summer, and it is used because it gives a clean, intense heat. If you want to use charcoal, and energy efficiency is important to you, then buying imported coals is definitely out. Why is it not a good idea? Because of unsustainable and unverifiable production, transport issues and because it is better to support your own, local economy. We provide charcoal for our Ecolodge guests but we produce our own in a homemade kiln, as described later in this chapter. Imported charcoal is also inferior in quality compared to the home-produced article as it takes much longer to heat up and produces less heat. Charcoal absorbs water very easily, so you will find bags of imported charcoal in the big stores weigh much more even though they are smaller than bags of locally-produced charcoal.

Charcoal is the remnant of wood which has been heated but not burnt. As the wood is heated it gives off its hydrocarbon gas and other biotic evaporatives. Having lost all of these gases and oils the charcoal can then burn cleanly.

Because it burns cleanly it is ideal for cooking, but charcoal has an insatiable appetite for oxygen and produces plenty of carbon dioxide and carbon monoxide, which are both poisons and have soporific properties. In the 'background' chapter (see page 13) I explained how wood burns, and charcoal is the third phase described there. I also covered the fact that half of the energy in firewood is produced in the burning of the gases, and so producing charcoal by burning some wood and the gases from the potential charcoal is quite energy inefficient. The reasons for charcoal production have to be good otherwise they would be outweighed by the energy inefficiency involved.

charcoal production

Producing charcoal is one of those things where it is easier to see the process than describe it. The basic idea is to roast split firewood in an oxygen-depleted, enclosed space to drive off all the evaporatives. We made a charcoal kiln out of a galvanised, industrial, hot-water cylinder that is about a metre in diameter. It is a tube, about a metre and a quarter high and is open at both ends. The top has a lid that fits inside and sits on a ledge about five centimetres below the top of the tube. The bottom is open to the ground and the tube sits on four manifolds made from box-section steel. These manifolds act either as chimneys or as air inlets. Look at the photos and you will get the idea (see fig. 13). As air inlets they introduce air into the centre of the fire at the base of the kiln, and as chimneys they create draught to keep the fire going and remove smoke and evaporatives that don't burn. The difference between the two depends entirely on whether we make it a chimney by adding on a chimney pipe and blocking up the external open end with soil or leaving the whole thing open as an air inlet.

fig. 13: charcoal kiln parts

setting up a burn

The kiln sits on the four manifolds and this leaves gaps under the kiln between the manifolds. The kiln is loaded first with paper and kindling, on top of which goes dry, split firewood that is all of a similar size: between five and eight centimetres thick. This is the 'charge' of wood. The fire is lit and slowly burns to an inferno. During our last burn the smoke was intense until the heat built up enough to ignite the driven-off gases, which then burnt with an almost clear flame. At this stage the lid has not been put on the kiln. The whole thing has to get really hot and the charge of wood will subside slightly. The time to put the lid on is usually about five minutes after you think it should be: as I say it has to get really hot. The built-up heat is what keeps the whole process going at this stage and all the charge should be soaking up the heat whilst charring on the outside. When the time seems right, slide the top on and, if it is right, the gases should fluff out of the bottom. By 'fluff' I mean be expelled but not quite explode. The recess between the lid and the kiln side is then sealed with soil, sand, ashes etc. (see fig. 14). This keeps the air out and creates the oxygen-depleted, enclosed space for the wood to be roasted by the fire below, but it cannot burn because of the lack of oxygen. At this point all four manifolds have a chimney inserted to keep the fire going and the heat intense, the air for combustion is drawn in through the gaps between the manifolds. The smoke should pluther, a term used for billowing with some velocity, out of the chimneys – which is why it's not a good idea to make charcoal in the back garden.

Once the smoke reduces, the gaps between the manifolds are blocked up and two chimneys are removed to change those manifolds into air inlets. When the smoke from the two remaining chimneys turns a hazy blue colour then it is time to swap the chimneys and air inlets round, and the chimneys in the new position will continue to smoke. When these then turn hazy blue, the chimneys are removed and all the manifolds are blocked up with soil and packed down tight. The idea is to prevent any air getting into the kiln to make sure the fire goes out. The kiln is left to cool for forty-eight hours before opening. If it is opened too soon it can re-ignite and this can happen very slowly after the charcoal has been bagged. What happens this is just a corner of a piece starts to turn white and in no time at all the whole lot is roasting hot.

fig. 14: charcoal burn in progress

Some of the wood may not have turned totally to charcoal, if this is the case it is easy to identify because it will be heavier and will not break up with a blow against the side of the kiln. These pieces, called brands, can be bagged up and kept back to use in another charcoal burn.

Charcoal made by this method burns very hot and so you do not have to use huge amounts, it is also local and has not been imported and travelled half-way round the world, and so it is altogether more sustainable. If you choose to buy charcoal rather than make your own, make sure it is locally-produced and not imported. This is another example of how we can individually make an impact on local economies, sustainable production, woodland and habitat management, and reduction of transport.

stove design

Let's think about stove components, all of which are essential for efficient burning.

stove door

This is used for loading fresh fuel and removing ashes. The seal around the door is essential for efficient control of air into the stove. A leaking door allows air into the stove even when the air intakes are closed and can lead to excess heat being produced when it is not required.

primary air intake

The primary air intake supplies oxygen to the heart of the fire to build up heat and keep the charcoal burning, see fig. 18: box stove (page 64).

secondary intake

This intake is designed to add air to the evolved gases while they are still in the heated zone, see fig. 18: box stove (page 64). The air and hot gases mix to give efficient, clean burning. The baffle helps to achieve good mixing while the gas/air mix is still in the hottest part of the fire.

chimney damper

This damper can be set to reduce the draught of the chimney, which reduces gas velocities and gives time for the secondary air intake to mix with the gases and burn while still in the heated zone. The damper also has the effect of increasing heat output by reducing losses up the chimney.

baffle

The baffle is found inside the stove and increases the length of the flue gas path before it reaches the chimney damper. It also increases turbulence in the gases and hence improves mixing of secondary air. In stoves without a boiler the baffle can heat up and so reduce the loss of heat when mixing hot gases and cooler air from the air intake. When a stove has a boiler attached the baffle can be part of the boiler, which increases transference of heat to the water, but also reduces the gas/air mix temperature. This reduction means that efficient, clean burning is more difficult to achieve.

chimney

This is a very important part of the stove: so important that it has its own chapter (see page 67).

ash grate

The grate at the bottom of a stove is only for coal burning, which needs air to be directed under and through the fire. Multi-fuel stoves can have either a pair of sliding grates that can be closed for wood burning, or a grate with narrow slits that will bridge with ash when burning wood.

Wood burns best when burning on a bed of ashes. These ashes keep the heat in and so keep the heart of the fire as hot as possible for the given stove settings.

If the sliding grate method is used then it is natural for the ash to build up, however the narrow slits of a single grate will soon bridge with ash and give the same overall result. The sliding grates are almost impossible to move without cleaning out the stove, and so it seems to me that little benefit can be gained by this design. The 1980s' Woodwarm we have has this design, but the grates have always been left in the open position so that we are able to clean out some ashes through the grate while the stove is still running. This is particularly relevant once your wood burning skills and firewood processing have attained a level of competence so that the stove is kept burning for weeks on end.

log grate

This is found on models where the air inlet is low down on the doors. It is just inside the doors and, in effect, is a vertical grate, only a few inches high, that stretches the width of the door opening. It has two functions: firstly, if the doors are left open to give a good view of the fire, then the log grate retains the logs and secondly, as the only air inlets are low down in the doors, the air moves through the inlets to an area just in front of the log grate. From this point the air can either pass through the log grate into the burning logs, or up between the grate and the doors to mix with the gases and so burn in the hottest part of the fire.

It's interesting to note that the ventilation in a room is increased when the stove doors are open, because a larger volume of air goes up the chimney. When coal-burning grates were the only source of heat, some houses were built with underfloor air supplies to the fire. This prevented air being drawn across the room and causing draughts, and reduced the amount of warm air that was sucked up the chimney. With modern houses and wood burning stoves this is not so much of a problem, but it is important to have an air supply and decent air circulation in a room with a stove. The fact that stoves draw far less air from the room than open fires means that the air supply can be drawn from across a room. In this way the air can circulate in the room, provide oxygen so we don't all fall asleep and also provide an air supply to the stove.

back boiler

Back boilers are discussed in more detail in 'water heating systems' (see page 77), but in addition it's worth mentioning that many modern open fires have back boilers, and many are designed to run central heating because they have a wrap-around design (see page 77). If you are living in a situation where you do not want to interfere with the structure of the building, or go to the expense of ripping out the old fireplace and installing a stove and liner, then there is an alternative which involves making a stove front to fit in the fireplace. A friend of mine made a stove from one of these modern wrap-around back boilers which he dragged out of a skip. It is only really cost-effective if you make it yourself and, as with any installation of this nature, some experimentation may be needed to get things right. I suppose you could get a local blacksmith or agricultural steel workshop to make something to an approximate design. I've included some photos to give some idea of the whole process, but it is based on the door design shown in the 'home stove building' chapter (see page 117).

In this example the boiler is a full wrap-around with a heating flue through the back of the boiler. There is a damper plate, which slides over the top of the boiler to cut off this rear flue and is used to direct how much heat goes to the boiler or to the room. The damper had to be modified to reduce the flow of gases up the chimney because the stove front reduced the flow of air from the room into the fire, made burning more efficient and reduced heat loss from the room up the chimney (see figs. 15-17).

fig. 15: fireplace conversion with door open

The final design is up to you but bear in mind that there should still be some way of making regular chimney sweeping possible.

fig. 16: fireplace conversion showing door hinges

fig. 17: fireplace conversion with door closed

types of stove

I am going to go through some basic designs to give you an overall idea of stove design but I will not include all the variations of the designs and no doubt, in time, other designs will become available. One of the major advantages of using a stove rather than an open fire is that the whole body of the stove gets hot and radiates heat into the room which would otherwise be lost up the chimney. All the designs described below do this but they each have their own characteristics that make them suitable for particular circumstances.

box stove

The basic principle of a box stove is that the primary and secondary air intakes are directly opposite the relative burning areas. The air controls are in the door, which has a double skin. The inner skin is drilled with a series of holes, which let the air into the combustion area. This series of holes is particularly relevant to the secondary air, because it encourages the mixing of burning gases and air to give clean, efficient burning. The double skin also has the effect of slightly warming the incoming air. The logs are loaded from one end and slowly burn from that end back down the length of the box. The drawing will make things clearer than me trying to describe it in detail (see fig. 18).

This type of stove is easy to make from bits of scrap from round the back of the workshop, with an angle grinder and arc welder. As all home-builder/inventors are fully aware, it's important to have lots of off-cuts of steel and bits of junk lying about behind the workshop. For one reason, it's a ready supply of odd bits that will do the job at hand and, for another, you have to have somewhere for the nettles to grow.

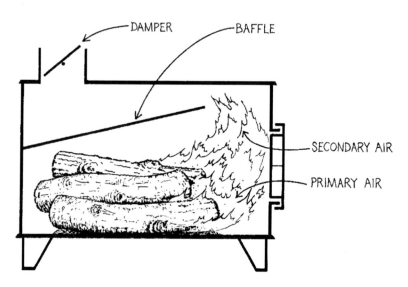

fig. 18: box stove

front loading stove

This is the type of stove that many people recognise as a wood burner. There are usually two doors at the front with glass panels, unless the glass has been broken and been replaced by bits of steel plate. The top of the stove comprises a steel hood, which contains the internal baffle and the connection to the chimney. The log grate keeps the fire away from the doors, and gives sufficient space for the passage of air from the door-mounted air inlets. Many of these stoves have an ash grate and are multi-fuel stoves, see the section on ash grates (see page 58). This type of stove can be used with the doors open to give radiant heat direct from the fire but this will increase the volume of air going up the chimney. The increase in air passage drags air out of the room and takes some of the heat with it. It is well known that open fires are very inefficient and using a stove with the doors open just makes it into an open fire with all the associated inefficiencies.

Clearview stoves

These stoves have a different air inlet system to the other designs I have described. The air is introduced through a narrow slot at the top of the doors: this slot is the full width of the doors and is about 8mm wide. The air is drawn down the inside of the door area and prevents the combustion gases from contacting the door, which prevents any deposits on the glass. This type of air inlet is used in conjunction with a modified baffle instead of the standard baffle. This baffle design prevents the inlet air being drawn directly up the chimney from the inlet and bypassing the fire (see fig. 19 & 20). This design of stove is popular, and gives better mixing of the combustion gases with the air, to give a cleaner and more efficient burn. Some of these stoves have an extra air inlet in the doors for lighting, and a thermostatic air inlet, which closes as the stove reaches a preset temperature.

fig. 19: Clearview air inlet

The stove in the pictures came to us via my mum because it had been stored in her garage for quite some time. Its burn characteristics are very user-friendly, insomuch as the adjustable thermostatic air inlet at the rear helps with lighting and keeping a low, even temperature. The Clearview air inlets at the top of the stove, just above the doors, keep the wood burning evenly, but it is controllable. The thermostatic air inlet adjusts the overall temperature, while the Clearview vents keep the wood burning steadily. It is now installed in my office and requires little attention once large blocks of wood have been added. I think it is grand.

fig. 20: Clearview air inlet (2)

Understanding these basic stove formats should be of help to the potential buyer of a new or second hand stove and will also be of use for reference when we get to 'home stove building' chapter (see page 117).

chimneys

Hot air rises, as we all know, and this is why it is difficult to heat large, high-vaulted buildings. All the warm air rises to the ceiling leaving cooler air near the floor. So, imagine a bonfire of dry, garden waste burning on a still day. The intense heat produced goes straight up but is not easy to see. Now, add a small amount of green material. The column of smoke will become visible and goes straight up until the heat is lost. Enclose this column of smoke in a tube and we have a chimney that directs the smoke and prevents some of the heat loss. The retention of heat in the chimney is important for several reasons. Firstly, the retained heat prevents water vapour and tars from condensing on the inside, and secondly it's the heat rising that causes the chimney to draw the smoke upwards. Remove the heat and you loose the draught and get a room full of smoke.

The chimney is just as important as the stove itself. Its first purpose is to get rid of the products of combustion, so we don't have red eyes and bad coughs, but beyond this it affects how efficiently the stove burns and how much heat is retained within the building. To understand the possible problems a bit better let's look at some worst-case scenario chimneys.

Take an open fire with a large chimney, as in a castle-type, baronial hall. The fire is in a large open-mouthed fireplace and air can bypass the fire and go straight up the chimney. The first thing one will notice is that the fire can be difficult to light, and it takes some time for the chimney to start sucking the smoke away from the fire. This sucking is called 'draught' and is an important feature of a correctly-designed chimney. Once the fire is burning well, the air that bypasses the fire has already been warmed in the room and this heat is lost as it goes up the chimney. Bypassing the fire also reduces the mixing of air and gases within the fire, and so the fire has to be kept 'roaring away' (burning rapidly) to draw air into itself to keep burning. As we have already discussed (see page 17) a rapidly-burning fire loses a good proportion of its heat up the chimney, because of the rapid movement of air and gases. The final problem with this situation is the large chimney itself which, being large, allows the gases to cool rapidly. This has two effects: reducing the upward movement or draught and providing ideal conditions for deposits from the smoke to form on the inside of the chimney – more about this shortly.

The second worst-case scenario is that of a large stove with a stovepipe or chimney liner that is too small. The stove will light reasonably easily but will have a tendency to leak smoke. Once the heat has built up the fire is fine, but it will fluff smoke out of the doors when they are opened, and the whole thing will have a lethargic feel, as it cannot really attain its full potential. This is because the flue-gas velocity is way below the optimum. This can also mean that tars will be deposited if the temperature is allowed to fall, as can easily happen with a stove that is difficult to regulate, so that it is either roaring away or nearly going out.

chimney deposits

In 'background', under phase 1 of 'how wood burns' (see page 17) there is a warning about deposits in chimneys or flues. Here's the reasoning behind the warning: if the flue gases – gases produced by the stove and now in the chimney – have high moisture content, then they will be relatively cool and tars will be deposited out of the smoke onto the chimney walls. These deposits can vary in structure depending on how well the fire burns and how well your wood has been seasoned. So, as a rule of thumb:

- the fuel should be dry so that the hydrocarbon tars in the evolved gases are burnt in the fire and not emitted into the chimney
- the flue gases should be hot enough to rise up the chimney rapidly.
- the chimney should be large enough to accommodate the gases without restriction.
- the chimney should not be too large to lose too much heat and reduce flue gas velocity.

chimney design

As already described, the flue should be of the right size: commonly between thirteen and twenty centimetres in diameter.

The construction should be designed to give the minimum number of bends; bends act as a restriction and slow down the gas velocity. There should also be a means of access to clean out the flue as part of regular maintenance, and sometimes provision of a soot trap at the base of the vertical stack (see fig. 21). There are all sorts of materials used in flue construction: from pot liners to double-wall, insulated stainless-steel flue sections. The choice of material depends on whether it will be used in a new building or fitted to an existing chimney.

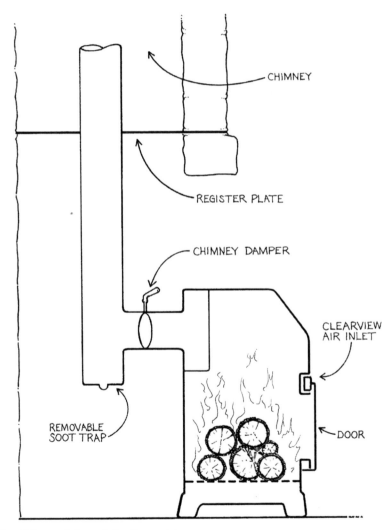

fig. 21: register plate & soot trap

new building

The choice of construction materials is determined by the materials used for the rest of the building, and the overall design. When we built the Ecolodge, which is constructed from locally grown timber, the obvious choice was a double-insulated, stainless-steel flue (see fig. 23). We did not want to build a large masonry structure with the associated large, concrete footings. The double-wall, insulated, stainless-steel flue was lightweight, took up a minimum of space, was easy to fit, and, because of the double-wall and insulation, prevented heat transfer to the timber structure within the roof space and reduced flue-gas temperature loss.

If a new build is in brick then pot liners in the new brick chimney may be more appropriate. These are terracotta tubes, about thirty centimetres long, that are built into the chimney to create a smooth, easy to clean internal surface. The material choice is dependent on its application and by agreement with the local building-control officer.

All flue construction should conform to building regulations, and anyone in the UK can phone their local building-control officer for advice, see 'resources' (page 141).

fig. 22: a stainless steel, double-wall insulated flue used for safety reasons and to limit loss of flue gas temperature

retrofit

In many older buildings the stove is fitted to an existing chimney and this can cause some problems. The chimney may not be structurally sound, or may have cracks or holes somewhere along its length. It is common for chimney stacks to weather and become loose, especially in pre-1930s' buildings. If there are structural faults then get them rectified before the stove is installed. Many chimneys are too big for efficient, long-term wood-burning use, and it is best to fit a liner in this situation (see page 75). In fact a liner is now required to comply with building regulations for any new installation.

The old way of fitting stoves into an existing chimney is to fit a flue pipe out of the back of the stove and into the chimney through a register plate. A register plate is just a metal plate that seals off the chimney and the pipe of the stove is fitted tightly through it (see fig. 21). Register plates are usually made of steel as they have to be very fireproof and able to resist melting if there is a chimney fire. The top of the register plate needs to be cleaned regularly to prevent the build up of flammable material which can cause a chimney fire. Access to the area above the register plate is required for the cleaning; this can be provided by building a small access door into the brickwork at the right level: these are commonly called 'soot doors'.

Most of the existing chimney is then used as a flue but this can lead to problems. For example, a friend of mine fitted a Rayburn cooker range to an existing flue. The flue was dead straight, well-constructed and in good condition. He fitted a flue pipe out of the back of the Rayburn smoke box and into the chimney through a register plate but had problems as the stove pluthered (billowed out quite fast) smoke out when the door was opened. He decided that the problem was caused by the pipe going horizontally out of the smoke box and so fitted another stove pipe out of the top of the smoke box and into the side of the chimney above the register plate.

Unfortunately he forgot to block up the first horizontal stove pipe. When he lit the stove again the new vertical flue warmed up and the old horizontal flue became colder. This caused an extreme case of 'downdraught' so that the column of smoke was spiralling up the centre of the chimney and cold air was being sucked down into each of the four corners of the chimney. In essence the cold air was spiralling down into the corners of the chimney and then being drawn back up in the warm flue gases. This, effectively, reduced the draught because air was not being sucked through the fire, but being sucked from the downdraught.

This showed up very clearly in this case because my friend had two stovepipes connected, but it is a phenomenon that can happen anyway with large, square chimneys (see fig. 23). Once his horizontal pipe was blocked up the stove behaved in a more manageable way, but there was always an issue with downdraught.

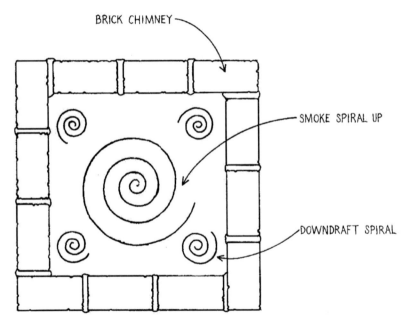

fig. 23: smoke spiral

downdraught

This is the movement of cold air down a chimney creating loss of draught and a room full of smoke. There aren't any hard and fast rules to apply to solve the problem, only a variety of things to try. As shown in the previous section, accident and observation can help identify the causes of downdraught and trial and error have to be employed to eliminate them.

wind

The chimney should be higher than anything else in the area. If there are tall trees or taller buildings upwind of your chimney, then the wind can curl over these obstacles and create an area of high pressure around the chimney stack. This pressure can be equal to, or greater than, the upward pressure, or draught, of the chimney. This causes smoke to puff out of your stove into the room. As well as this the wind passes over the roof and can create an area of low pressure in the lee, downwind, of the house. Opening a door or window on this leeward side means that air is sucked out of the house into the low-pressure area. This air is replaced by an air/smoke mixture from the high-pressure area above the chimney, and so smoke is drawn into the house from the chimney.

To test to see if this problem is causing your stove to smoke, keep all the windows and doors closed, and then open a window on the windward side of the house as this can balance the pressures; the stove should then stop smoking. Unfortunately it is seldom that simple, because the downdraught/smoking problem may have more than one cause. A process of elimination and experiment is definitely called for.

Here's an example of a more complex situation where a stove smoked when the wind was in a particular direction. There were no obstacles upwind to create an area of high pressure. The chimney stack in question was on the windward side and about half way up the slope of a roof, but it was the same height as other chimneys on the house, which did not smoke. All chimneys were about four feet above the ridge of the roof.

As an experiment we attached some little flags and streamers to areas of the chimney and roof area. The streamers showed that once the wind got over a particular speed, as it hit the roof it caused a localised high-pressure area. The experiment was continued by fitting a length of stovepipe to the chimney pot. This extended the top of the chimney outside the high-pressure area and stopped the smoking. The pipe is still there two years later waiting for inspiration and a permanent solution but we did find the cause of the problem.

liners

I am not going to give detailed information about liners as there are all sorts of products on the market and suppliers can give you precise specifications for your circumstances. The main thing is that you know that liners can be fitted and it is not as difficult as some may like to make out.

Liners for chimneys should give a smooth, internal face for easy sweeping. Liners for straight chimneys are not a problem to fit, but it is more difficult for flues with bends. Double-wall insulated, flexible, stainless steel liners are available which are suitable for this purpose. The inner face is slightly ridged to give flexibility and these ridges should face downwards so as not to produce a series of little shelves facing upwards that will collect debris. It is often necessary to cut access holes in a brick chimney to feed the new liner round the bends and these can then be bricked up and re-plastered.

A liner is now required to comply with building regulations for any new installation.

smell

If the room where your stove is smells of smoke or tar then the most likely culprit is the quality of the firewood used: the moisture content may consistently be too high. In the worst cases the condensed tar can run down the chimney and exude out of chimney joints. It is black, horrible, sticky goo, which we have previously explained the causes of.

If firewood quality is not the problem and the stove is not lit then there could be a temporary downdraught, which would disappear if the stove was alight. In this case just close the stove vents and the chimney damper.

If the stove smells of smoke just after it is lit, then this could be temporary until the chimney warms up, or you may have the stove settings wrong. The air inlets could be open too wide for the chimney damper setting. It could also be a downdraught problem as previously covered.

chimney sweeping

If a stove is in regular use the chimney must be swept at least once a year. You can get someone to do this for you; a local chimney sweep will have all the right kit and be able to make a clean job of it. Or you can do it yourself. If you have a set of drain rods a flue brush head can be purchased, from a hardware shop or building merchant, to fit on to it. If you decide to buy drain rods for the purpose they come in one-metre sections and have brass connectors that screw together. So you will need to work out how long the flue is and buy the right amount.

Some flue systems have cleaning access points and soot traps, which makes cleaning the chimney from ground level easier. If they aren't there you may have to dismantle parts of the flue pipe. Depending on how difficult the flue cleaning points are to access the chimney may have to be cleaned from above, which could mean going up on the roof. If you are fitting a flue it is beneficial to think about these sorts of things during the design process and to make the system as easy to clean as possible.

If your stove has been installed using the register plate method there should be some access so that you can remove the debris that falls on to the registry plate. A build up of these flue-gas deposits is a common reason why chimney fires occur. Before cleaning from either direction, make sure you cover furniture and carpets with dust sheets to protect them from the soot that will inevitably be produced.

Cleaning is no problem if the combustion in the stove is correct. If, however, your fires have been producing tars and creosote from unseasoned wood then things can get tricky. The inside of the chimney or flue will be coated with either sticky goo or hard varnish-type deposits which are impossible to remove. There are products available that can be burnt in the fire and are intended to remove these deposits. I have never used them myself so do not know how effective they are. The best thing to do is burn well-seasoned firewood with a bright flame and this will not become a problem.

water heating systems

In this chapter the intention is to outline a simple hot water system. The information given here will not be sufficient for the installation of a system. It will, however, provide the basics so that further reading will make a lot more sense.

Many stoves have a boiler fitted inside them to provide hot water. These can be referred to as 'back boilers' because they are usually at the back of the fire. There are other types of boiler called 'wrap around' boilers which, as the name implies, are at the back, sides and top of the fire. All stoves have a baffle, which directs the burning gas and air in a circuitous route. This baffle can also be part of the boiler. The boiler takes heat away from the fire and transfers it to the water contained within.

Now, as we have mentioned before, hot air rises and so does hot water. Within a body of water the warmest water will be at the top. The boiler has two or more connections but for the moment, to keep things simple, we will imagine that there are only two.

Of these two connections one is higher than the other. The highest is the hot connection and is commonly called the 'flow'. The lower connection is the cooler of the two as it is at the bottom of the boiler, and is called the 'return', as water returns to the boiler.

The boiler is connected to a hot tank so that the hot water can be stored. It is common for hot tanks to have a large copper coil fitted inside them, which acts as a heat exchanger. The 'flow' is connected to the top coil connection, and the 'return' to the bottom. There is also a small tank, normally in the loft, that keeps the whole system full of water (see fig. 24). We will add other bits to this system later but this is the simplest version.

The water in the boiler and hot tank coil just keeps circulating and moving heat from one place, the boiler, to another, the hot tank. No water is lost from this system beyond that of evaporation.

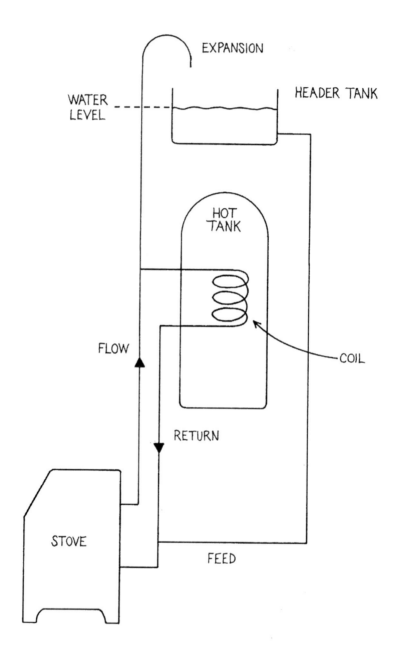

EXPANSION

HEADER TANK

WATER
LEVEL

HOT
TANK

COIL

FLOW

RETURN

STOVE

FEED

fig. 24: simple hot water system

So let's just think this through. The stove is lit and heat is transferred through the boiler into the water. As the water warms up it starts to rise and begins to move up the flow pipe. This movement causes water in the return to move into the boiler. The hot water enters the coil and, as it transfers heat to the water within the hot tank, it starts to descend through the coils towards the return. As long as the stove is giving out heat and the water is hotter than the water in the hot tank then this circulation will continue. It is only the heat that makes it move, no pumps are needed. This type of system can be used in many ways and is called a gravity system. The pipes connecting the hot tank and the stove should be 28mm diameter to enable the water to flow easily. The flow should always be on an upward incline, or straight up, and be as short as possible, and the return should follow it so that the water can return as quickly as possible with the minimum of bends.

Water expands as it heats up and, if heated within a confined space, can create extremely high pressure. For this reason there must be some way for the water in the boiler/coil system to expand without causing pressure. In the case of a simple, vented system, an expansion pipe is added which rises above the header tank and hooks over it. In this way any water forced through the expansion pipe by the act of water boiling in the boiler will be returned to the system. This is where problems with boiling water can start, as mentioned in 'beginners' guide to stove use' (see page 103).

hot water tank

Hot water for domestic use is drawn off from the top of the hot water tank. It must come from the top as the hottest water is always found at the top of the tank. Water of differing temperatures doesn't mix that easily so, as hot water is drawn off from the top of the tank and cold water replaces it at the bottom, the cold doesn't immediately mix with the hotter water to reduce the temperature. The hot water rises and the cold water sinks and so the various temperatures stratify into different layers. If it wasn't for this property, hot tanks wouldn't work the way we are used to.

All new tanks are now pre-lagged with some form of urethane foam but old tanks need to be lagged.

heat link

Our system, as so far described, has heat from the boiler going into a hot tank which, if left to its own devices, could reach boiling point. The solution to this is to be able to divert heat away somewhere and for that heat to be useful.

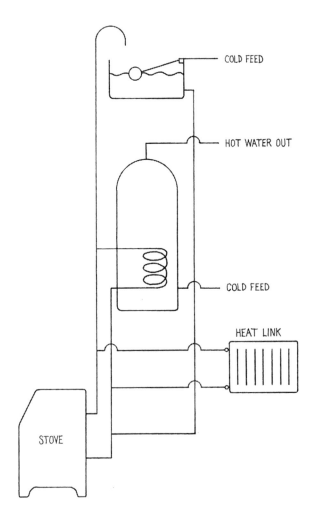

fig. 25: heat link

If you are like me you will remember bathrooms that were damp and freezing in winter; that never has to happen if the heat dump from the boiler is a radiator in the bathroom. Basically, a single radiator can be plumbed across the flow and return wherever it is convenient and when the stove is lit all the heat goes up the 28mm pipes to the hot tank, but if the heat differential between the hot tank and the boiler is reduced and the water velocity slows, then hot water will divert to the heat link (see fig. 25). The result: no boiling water, dry towels, and a delightful, dry, warm bathroom.

central heating

So far I have explained a basic hot water system based on a solid fuel boiler. Central heating radiators work in a similar way to the heat link. A connection can be taken from the flow pipe above the hot tank connection, which then feeds the radiator system (see fig. 26). The return from this radiator circuit is connected to the return near the stove wherever it is convenient. Some stoves have more than two tappings into the boiler; these extra connections can be used for direct connection to the central heating radiators.

Water can be circulated around the radiators by two basic methods: by using a pump or gravity. A pump can be attached to thermostats and time clocks so that radiators can heat up at pre-determined times, and heat can reach the furthest and most-difficult-to-heat points of the house.

Using gravity works in the same way as the primary flow and return to the hot tank. Large, 28mm pipes are required and the pipe routes need to be carefully constructed to prevent air-locks and restrictions to the flow of water, because there is no pump to help the water move and dislodge those difficult bits of trapped air. For the same reason, gravity systems also need a different style of radiator valve from pump-driven systems. Normal radiator valves have a small-diameter pipe sticking up inside the casting, over which a rubber, tap-washer-type seal is pushed to cut off the flow of water. In a gravity system the restriction to flow in standard valves is unacceptable and so a full-bore radiator valve is required, which allows water to flow straight through. These have to be ordered from a plumbing stockist and cost a bit more

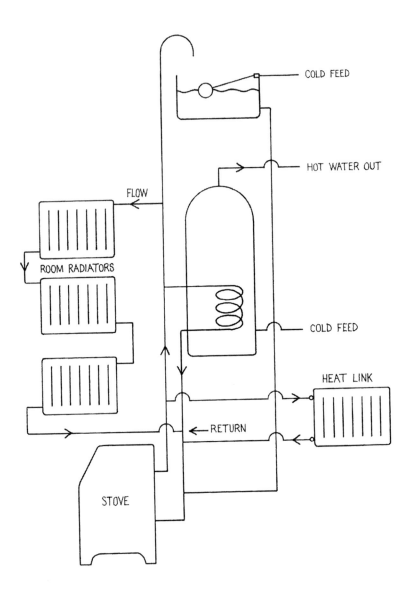

fig. 26: central heating

than ordinary valves. They are also needed for the heat link radiator in a normal, pumped system. In the Ecolodge the heating works using gravity and the system is tuned to give preference to hot water before space heating. I am pleased with the results, as it is a gentle type of heat. I have included an illustration of a simple system to give you the basic idea (see fig. 26), but for further details on in-depth plumbing and information you will need to do your own research.

All central heating systems contain a mixture of materials: copper, brass, and steel. The water filling the system, which is constantly moving, contains oxygen and can start corrosion. If the system has a leak then the water is being constantly topped up with oxygen-rich water and corrosion will continue. The other major constituent found in water in many areas of Britain is mineral, particularly calcium. This is deposited on the inside of pipes, as lime scale, particularly when the water is very hot. These deposits reduce the flow of water until the pipes are blocked. There are additives available to add to the boiler water that prevent corrosion and lime scale.

solar water heating

Heating domestic water with solar panels makes sense and can fit into a wood heating system very well. The system works on the principle that the stove runs most of the time in winter and so a large store of hot water is not required, whereas in summer the solar panels capture the heat when it is available and store as much as possible.

The general idea is to install a larger hot-water cylinder than usual, that has two heat-exchanging coils in it. The coils are one above the other and the solar panels are connected to the lower coil. The coil at the bottom heats the water, which then rises to the top of the tank, or to a position where the water is the same temperature. In this way the coil at the bottom can heat the whole of the tank and make the most of the available solar heat.

There is another method that can be used when it is difficult to fit a larger tank, or the tank is in the wrong place for solar connections: this is to have a secondary tank in line with the primary tank. This way solar energy heats the water, which then feeds into the primary tank on demand. If the water from the solar tank is hotter than the primary tank then it will rise to the top of the tank, and so be readily available (see fig. 27).

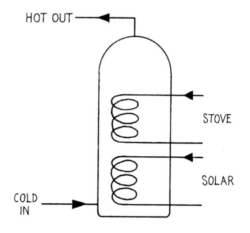

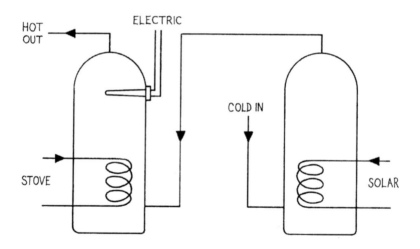

fig. 27: wood fuel and solar hot water connections

automated systems

The idea of automated wood-fuel systems, which are also referred to as 'biomass boilers', is to have reduced manual interaction between boiler and homeowner. Some of these systems can almost behave like a gas or oil boiler.

The fuels used for automated systems are mainly wood chip or wood pellet. This chapter is an overview of the basic ideas and no specific boiler details are given. When considering installing one of these systems it is important to deal with a company that a) can give backup, b) is local, if possible, and c) keeps a stock of spare parts.

wood chip

Wood chip is exactly what the name suggests: woody material that has been fed through a wood chipper (see fig. 28). For quality wood chip the material needs to be screened to remove dust and chips that are too large – I will explain why later on. The maximum size of chip depends on the size and specification of the boiler and the chip-feed systems, as seen with the Alcon boiler feed that will take up to 7cm chips, (see fig. 32). The moisture content is also important, for two main reasons. Firstly, if the chip moisture is too high (above thirty per cent) then the boiler will be unreliable and will continually be going out. The second reason is the same as with ordinary firewood: the system needs efficient heat production to avoid creating chimney problems. Ideally the moisture content should be close to the fibre saturation point (twenty-five per cent) but twenty per cent is even better. These moisture and quality issues mean that it is not possible to fell a tree, chip it straightaway and expect it to produce heat in a biomass boiler. If you do the boiler will go out, the chip feed system will clog up, and the chips will rapidly begin to rot and stick together in a gooey mass.

fig. 28: wood chip

pellet

These consist of small particles of biomass that are extruded at high pressure to create a fairly standard-sized burning pellet (see fig. 29). The pellets produced are around 8mm in diameter and vary in length, up to about 20mm; 6mm pellets are the easiest for a small auger to handle. As the pressure forces the material through the extrusion head a large amount of heat is produced, which softens the lignin within the wood and, as the pellets cool, the lignin hardens and glues the pellet together. The process creates a standard-sized and clean product, which can be handled with ease in automated systems.

fig. 29: wood pellets

This is, of course, ideal for urban dwellings where compactness, cleanliness and convenience are compatible with a modern lifestyle.

feed systems

The fuel is usually fed into the combustion chamber by means of an auger. An auger is a very coarse screw thread that turns in a tube. Material introduced into the gaps between the threads naturally moves along the tube as the thread rotates. It's a lot like an Archimedean screw except there the tube doesn't turn. The quality and robustness of this auger dictates to some extent whether the system can deal with chip and pellet, or only pellet. It takes a robust auger to deal with wood chip.

The auger is driven by a motor and moves fuel from the fuel hopper to the combustion chamber. The motor does not run all the time, but introduces fuel in regular bursts depending on heat output demand (see fig. 30).

fig. 30: feed auger and agitator

hoppers

In small units suitable for households the hopper is part of the stove assembly. This is filled manually several times a week. With larger boilers a huge, remote hopper is required that can be filled by lorry. Hoppers also contain an agitator that keeps the chips flowing instead of just removing the fuel from the centre of the hopper, which would happen if only gravity was relied on. The agitator is just a slowly rotating motor with bendy arms attached, that stirs up the fuel to keep it moving into the auger at the base of the hopper.

combustion systems

The fuel is either introduced by the auger to the fire from above or pushed up through the fire from below. Fully automated units are self-lighting, with heating coils to heat the fuel to a point where it ignites. A fan then encourages flames and the production of heat (see fig. 31). Other systems rely on the chips smouldering between heat demand cycles and they are manually lit if they go out. Wood chip and sawdust will smoulder for a long time. We produce plenty of sawdust, chainsaw chippings, and shavings. These are used in the compost toilet but, when supply overwhelms demand, we light a pile and it will burn steadily for days.

fig. 31: biomass flames

multi-fuel boilers

The design and robustness of the fuel-handling system dictate whether the unit can handle pellet only or pellet and chip. There are some boilers that are built to also act as log burners. The system has a chip/pellet burner attached to a log burning boiler. In this way the choice of fuel is increased to include logs. The

fig. 32: 70Kw Alcon biomass boiler

biomass burner fires heat into the log burning boiler and so transfers the heat to the water or the log burner is batch fed with logs to do the same job. Batch feeding boilers require more interaction with the homeowner, and the output is more variable.

I have been to see an Alcon boiler with a Gejs hopper and stoker system for pellet and wood chip. It was most impressive with a 70 kilowatt boiler feeding a district heating system on a farm. The buildings heated were the large, rambling farm house, the farm offices and three holiday cottages. Fig. 32 shows the boiler, which uses log, and chip or pellet. The stove can be fed with batches of logs and has an air-temperature thermostatic control. If the logs burn out then the chip/pellet burner will strike up and take over. This gives extra flexibility. The chip auger is impressive and runs in a square tube to prevent blockage by difficult-sized material. It will handle up to 7cm chips. This can be seen to the right of the boiler on the photo, the square box between boiler and auger is the burning head. Have a look at the picture and for details, see 'resources' (page 141) – the supplying company is Louth Tractors Ltd.

batch-burner boilers

All wood stoves are effectively batch burners, in that the stove is filled with a fresh supply, or batch, of logs at intervals depending on the amount of heat required. There are boilers for central heating that work on the same basic principle of batch-feeding fuel, but they are larger and the air intakes are automatically controlled. The most practical situation for them is in their own boiler house or room. These boilers are neatly divided into two main types depending on efficiency and firebox size.

The less efficient, and cheaper, type (see fig 33) has a large combustion chamber that is filled with firewood as required. The air inlet is thermostatically controlled to keep the water temperature above an adjustable baseline. These stoves, when in slumber mode, are inefficient and can produce hydrocarbon gases. This produces considerable quantities of visible, light or dark grey smoke from the chimney and can annoy the merry hell out of your neighbours, see 'clean air act' (page 108). Careful selection of firewood and using large pieces, can, however, enable the stove to be set so that a bright flame can burn off

fig. 33: basic batch boiler

some of the evolved gases without producing too much heat. Experience and commitment is required to get the best combustion, but these stoves will take up to one-and-a-half metre length logs of considerable diameter. The large firewood can stay in the stove for several days, with smaller material being stacked around it at intervals. One of the major side benefits of this type of boiler is that there is considerably less firewood processing

fig. 34: Alcon 30kw biomass batch burner, suitable for a medium-sized house

required than for ordinary stoves, but on the other hand large firewood takes much longer to season. It is interesting to note that oak logs of about 37cm in diameter can take more than five years to dry if they are not split, and even larger material seems to take forever.

The more efficient type of batch-fed boiler has a smaller combustion chamber, which is automatically fed with logs from a log compartment. This enables a smaller fire to keep burning with a bright flame so that all the evolved gases are burnt off. We already know that half of the potential heat from firewood can be obtained from burning these gases. These systems are necessarily more complicated and so more expensive. The firewood also needs careful processing to a smaller and more exact size than the previous type, but this does enable better seasoning and, as it is smaller, the firewood is easier to handle. It is also interesting to note that grant funding for installation is only available for these more efficient boilers, which makes sense.

fuel supply

These automated systems have been used for decades in other parts of the world. In the UK suppliers are just getting their act together, although that statement may well be out of date soon, but there is currently a fuel supply and demand problem. As more systems are fitted then the supply side will have to become more integrated. I can see traditional coal merchants changing their core business to accommodate these fuels; after all they have the yards and delivery systems that are ideal for pellet fuel, which comes in bags.

stove size

Every stove or boiler has its own output rating. The output is more reliable with this type of boiler because they use standard-sized fuels. Stove size can vary enormously though, from a pellet burner that is about the size of a large wood burner to units designed to heat a whole school. Pellet stoves are ideal for small properties because they will not be running full time, and so the flexibility, size, and self-starting are all important benefits. There are pellet burners with back boilers that will fit in an average front room and run the central heating system, and only need to be filled a couple of times a week depending on the amount of heat required and outside temperature.

district heating

On a visit to Kielder forest in Northumbria a few years ago I was introduced to the biomass district heating system that heats some of Kielder village. The standing, dead, dry trees from the surrounding forestry blocks are chipped and used as the fuel. The fuel store is the size of a large, double garage and chip is moved to a feed auger by a series of bars that move over the floor, which is known as a 'walking floor'. The large auger then feeds a smaller fuel store which the feeds the boiler. Each of the houses or units on the system has its own heat meter which enables individual billing for the energy used. One could look at the main heat supply as a boiler that supplies heat on demand. Such systems are much more efficient than separate boilers in each unit, are using fuel from the local area, contribute to the local economy and reduce fuel miles.

heat store

This is a large, insulated tank with a volume of between one and two thousand litres which is used to store hot water and can be used as an add-on to a domestic system. The extra capacity allows the fitting of a stove that is not large enough to cope with peak demand, but has sufficient output most of the time.

It works as follows: the central heating pumps come on at a particular time after which there would be a time delay as the boiler fires up from slumber mode and builds up heat. With the heat store in place hot water for the heating system is available on demand, and allows the smaller boiler to build up heat in its own time and then replaces the heat used from the heat store. A smaller boiler working a greater proportion of the time is more efficient that a big boiler only working occasionally.

what system is best for me?

From this chapter you can see there are several systems available. The choices you make will depend on the cost and complexity of the system, and the amount of operator time and maintenance you want to put into it. The following summary will give you and idea of how the systems compare.

The least expensive and inefficient batch burners require manual feeding at regular intervals, at least twice a day, but can burn a wide range of material. They tend to be quite mucky and emit large amounts of smoke at times.

The more efficient batch burners, when used with a heat store, only require lighting once a day, and need a slightly better fuel to attain best results and complete heating of the heat store in one burn. These stoves are considerably more expensive than the inefficient batch burners but are cleaner to use and emit less visible smoke.

The chip burner is more expensive than the efficient batch burner because it requires a chip store, feed auger, and burner head. It also needs specific fuels of a correct size and moisture content. There is far less maintenance required and it can provide a constant heat supply. Most systems use bought-in chip, which takes up a large amount of space; for example a twelve-tonne grain trailer will carry about three tonnes of wood chip at the right moisture content. Some boilers are designed to use a variety of fuels, in that they can be used as batch burners with the chip burning as a standby. So, for instance, if you forget to feed the boiler with a fresh charge of logs, the temperature of the boiler goes down and the chip burning head will then fire up and continue heat production until more logs are added. This type of installation is suitable for large houses and district heating.

Pellet burners can vary in size from a small room heater to large boilers. The pellet is more expensive than chip, but has a greater density and so takes up much less room. The hopper for the small, room-heater stoves only needs to be filled every three to four days and the pellets can be bought ready-bagged. Although pellets are much more expensive than the chip, the fuel is more reliable and easier to handle. These systems are ideal for use in towns, smokeless zones and minimum-operator situations. Small pellet stoves are more expensive than standard wood stoves but require less day-to-day attention and the fuel is easier to handle and store.

beginner's guide to selecting and installing a stove

As you can imagine the process of stove selection is fraught with complexity. It can, of course, be as simple as:
'Do you want this stove for ten pounds?'
'What will it heat?'
'We used to run all the radiators in the house, and it did that fine.'
'Right you are then, I'll have it.'

That little scenario covered some (cost, what it will heat and availability) but not all of the considerations. It missed out things like: what it looks like, what colour it is, what spares are available, and how efficient it is.

There are many considerations but I would suggest that the primary ones are:

- cost
- heat output
- design
- duration of burn

cost

You will know what your budget is and need to work within it. Try to avoid being influenced by too much marketing pressure – the best stove for your situation may not be the most expensive. Having said that, if you are careful in your research, you get what you pay for and quality shines through. New designs don't always work as well as is claimed, so avoid the marketing hype and work on real facts. If there are no verifiable facts about a stove then I would take any information with a pinch of cynicism.

heat output

Stoves are all given a heat output rating by the manufacturer. In reality the actual heat output depends on timber type, size of logs and their moisture content and it is not possible to attain rated stove-output at all times. We must assume that any figures given by manufacturers are at best measured under perfect conditions. Keep this in mind when buying a stove and go for the larger rather than the smaller model if there is any doubt about which to choose. Making the right choice could mean the difference between spending all your time feeding a small stove burning at full belt to get enough heat, and having a larger stove burning steadily. This question links in with duration of burn and how this fits in with your lifestyle.

The best way to get an idea of how big the stove should be for the size of house and the number of radiators you want to run from it is to ask the dealer who is selling the stoves. This can be backed up with manufacturer's information and guidance. There are also websites that provide heat output calculators, see 'resources' (page 141).

Another thing to remember is that the more heat you produce the more timber you will use, so you have to make sure your woodsheds are large enough to store enough seasoned firewood.

If you decide you don't want to install the stove, water heating system or flue yourself, talk to your dealer. They may have their own installers or be able to recommend reliable, local companies to do the job for you.

design

This not only covers what the stove looks like, but the various features that make using the stove easy and flexible. These include:

- has it got a back-boiler, if you need one?
- how big is the back-boiler; is it just at the back or is it wrap-around?
- is it a Clearview stove, which will be more controllable and efficient than other types?

- does it have a thermostatically-controlled air intake?
- what is the average heat output into the room and to the water?
- is it easy to clean out, or does it look like a dirty, awkward job?
- is it multi-fuel?
- is there a choice between top and rear chimney-pipe outlets?
- has the stove got a hotplate, but do you lose the hotplate if you have the stovepipe fitted to the top?
- can you cook on it and does it have an oven?

It is a good idea to do some research and get to grips with the various features and how different manufacturers put them together. It is also beneficial to buy from a company that has been around for a while, whose spare parts are easily obtainable and will continue to be in the future.

second-hand stoves

It is possible to buy and install second-hand stoves, see 'resources' (page 141), which can be cost-effective and means you are recycling as well. If you decide to buy second hand always inspect the stove very carefully before you hand over any money.

Here are some of the things you need to think and ask about:

- whether the stove is still made and are parts available.
- whether the stove is the right size for what you need.
- if the price is really low and the stove needs some work, are you willing, and able, to repair it or would it be better to spend a bit more for something in better condition?

Many parts of the stove could be damaged so check all of the following carefully:

- is the fire string and are the seals around the doors intact? Remember, old stoves can have asbestos string so be very careful when handling or removing it.

- is the door glass intact and is replacement glass still available?
- are the door hinges and catches intact and in good order?
- are the castings intact and not cracked or burnt away?
- in a multi-fuel stove the grate could be burnt and melted.
- are there any areas of the stove body that seem to have been overheated? If they are they will be bent and twisted.
- if there is a boiler are the connecting bosses and threads in good order? If the stove has been used without the boiler connected to a water system, then you can almost guarantee that the boiler will leak and need replacing. In this case the firebox side of the boiler will generally be distorted or even burnt through.
- is the stovepipe boss with the stove and in good order, and is there a chimney damper with it? If the stove has been fitted badly in the past then there could be corrosion and cracking in the boss; this may have been caused by the expansion and contraction of fire cement.
- are the fire bricks intact and are replacements available if you need them?
- the baffles on many Clearview stoves are removable for cleaning. If you are looking at a Clearview stove check that the baffle is intact and in reasonable condition – it could even be missing.
- if the stove has been stored outside for any length of time it will have got wet. If that is the case then any bolts and fittings may have seized up. This can make things very complicated – there is nothing worse than having bolts that are threaded into a casing shearing off, and turning a five-minute job into a three-day nightmare.

installing a stove

If you're in a situation where it's freezing outside and there is no heating, then fitting a stove can be as quick and rough as needs be to obtain some heat. Imagine the situation of total energy collapse; in such a situation an old steel box from the scrap pile and a piece of pipe stuck through the roof would give the desired results. Luckily, for the moment, things haven't got that bad and so we have the luxury of taking our time and making things more permanent and efficient.

The position of the stove in a building depends on many factors that are as diverse as:

- whether the stove is a cooker as well as a boiler or room heater. If it is a cooker then one assumes it should be in the kitchen.
- whether the stove is attached to the water system or is it just a room heater?
- the position of existing chimneys and flues.
- whether a new chimney or flue is required.
- the proximity of the proposed or existing hot water cylinder.
- connections to a central heating system.
- the dynamics of the living space.
- fitting a flue.

Each house or living space has its own set of constraints but let's assume, for ease of explanation, that we have a choice to either use an existing chimney or fit a new flue. The fitting options for using and existing chimney are explained in the 'chimneys' chapter (see page 67), and the stove can sit within the existing reveal or sit on a new, fireproof base.

The construction of this base depends on the stove construction and the floor materials. If the floor is concrete, tiles or other such non-flammable material then there is no problem. If, however, there is a timber floor then the flammable material must be protected from the heat and possible ignition. For a stove that is raised up off the floor on legs or has an ash pan on the bottom, then the simple solution is a sheet of heat-resistant board with a sheet of 2mm steel on the top to give a good wearing surface. An edging strip is then required as a retainer for the fire area and to make a transition to the rest of the floor.

The other method, used years ago, was to cut out the floor, trim the floor joists and cast a concrete block within the wooden floor structure. This can then be tiled and trimmed round with a decorative timber edging making sure that all timber is well away from the heat. I don't recommend this method unless you are a carpenter, in which case you may well be familiar with the techniques of trimming joists and concrete formwork.

If a new flue is installed then the choice is between building a completely new chimney, which is quite a dramatic and expensive process or using double-wall, insulated, stainless steel, solid-fuel-grade flue sections. This is relatively straightforward, and the flue can go through an outside wall with the top above the ridge of the roof or through the internal roof space and out through the roof using a weatherproof-sealing fitting.

Which ever method is used one must always fit access doors for easy cleaning of the flues. It's a right royal pain having to take half the system apart on a regular basis as part of your routine maintenance schedule – as my mate Pete found out with his commercially-fitted stove.

The 'chimneys' chapter (see page 67), gives insight into design and avoiding downdraught problems. It is important for the top of the chimney to be above the roof ridge and so it is far more practical to mount an external flue on the gable end of a building, rather than going through the roof space, as otherwise a long section of unsupported flue would require a steel support structure.

There is a something called fire cement which is handy when fitting stove parts together but has its own drawbacks. It is a putty-like substance that is easily worked with the fingers and water soluble when in this state. This means that if your tub of cement has dried out a bit then you can just add a little water and rework it. This stuff can be used to seal joints in stoves and flues and fix firebricks in place – very useful. The cement is cured when the stove is lit and gentle heat is required initially. The downside of this material is that it expands slightly when curing and so you have to be careful with cast iron fittings – to prevent cracks in the iron. I try to avoid its use except for firebricks because it has a tendency over time to crack up (like most of us) with the constant expansion and contraction as the stove heats up and cools down.

There are snippets of information relevant to this subject scattered throughout the book but particularly in the 'chimneys' chapter (see page 67), and 'back boiler' section (see page 59).

beginner's guide to stove use

A few years ago we completed a long-term project and the Ecolodge was ready for guests. It's a holiday lodge that's run on just wood and wind power. What was surprising, although it really shouldn't have been, was that quite a few people had never used or lit a wood stove before. This is for the reasons I explained in 'background' (see page 13) at the beginning of the book, coupled with the fact that many houses in Britain have been built without chimneys: no doubt due to such factors as space, cost and the prevalence of gas central heating. So I have come to appreciate that using a wood stove for the first time is an unknown quantity for many people and can be an exciting experience. This section is intended to help those people get started.

preparation

Wood fires burn best on a bed of ash, which retains heat in the base of the fire. This means that it's not necessary to remove the ash religiously every day. Let it build up until the ash becomes a problem and reduces the log holding capacity of the firebox.

The first thing required to make a fire is kindling. I covered this at the end of 'buying and storing wood' (see page 35). People normally use more than is necessary, but if you have to split the kindling you soon learn to be economical with it. Paper or cardboard is ideal for starting the fire initially. Cardboard can be torn into narrow strips, but paper needs to be twisted to prevent it burning too quickly (see fig. 35). Use about five sheets.

Place the paper twists in the bottom of the firebox followed by no more than a dozen bits of kindling (see fig. 36). At this point make sure the chimney damper is wide open; the operating lever is normally in line with the stovepipe when it is.

Light the base of the paper (see fig. 37) and close the doors, making sure the air inlet is open. As the kindling starts to burn it will crackle and the flames will fill the stove. When it is burning

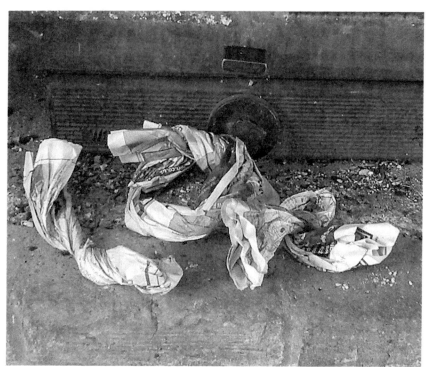

fig. 35: paper twists

well, (see fig. 38), add some larger pieces of wood. There is no point in putting large logs on at this stage as there is not enough heat to burn them. Split pieces up to about 7cm in diameter, like big kindling, are ideal. As the fire takes hold and the heat builds up the flames will start to be drawn up into the chimney and flue. This is not an immediate danger, and any risk depends on whether you are using a register plate and original brick chimney, where there could be deposits on the top of the register plate, or a flue with a liner. Flames roaring up the chimney can cause a chimney fire, but only if there are large amounts of tarry deposits in the chimney. As you are using dry firewood and have a well-designed, effective chimney you will be fine.

Once the fire is established the stove will give out heat to the room and into the water jacket. This is when careful use of the stove controls will give maximum efficiency. We have already covered wood burning in some detail (see page 17). The correct mixing of air supply and evolved hydrocarbon gases is essential at this stage. The challenge is to balance the loss of heat up the

fig. 36: kindling

chimney, with efficient burning and controlling how much heat is produced.

Here are some of the essential criteria:

- heat going up the chimney is good for draught and reduces the condensation of tarry deposits.
- it is essential to mix air and gases to burn off the tars.
- burning the gases gives fifty per cent of the available heat from the wood.
- when gases are burning well the flue gas velocity is too fast.
- the chimney damper, if fitted, is then used to reduce the flue gas velocity, but not to the extent that the fire smokes.
- reducing flue gas velocity increases heat output into the room.
- reducing flue gas velocity prevents flames being drawn up the chimney.
- reducing the air input will slow down the burning process.
- balancing the air input and chimney damper is better that just closing the air input.

I am sure you will get the hang of it quickly. The basic rule is: don't just light the stove and walk away. You need to observe what happens and get to know how your stove behaves.

fig. 37: lighting a fire

The things that can happen if you just light the stove and leave it are:

- the fire goes out due to using up all the available fuel or it never lights properly because of poor lighting materials.
- if there is sufficient fuel then the fire can burn too fiercely and create more heat than is required.
- the water in the boiler can boil which isn't in its self instantly dangerous, but it can have a detrimental effect on plastic header tanks and can create ideal conditions for the system to start 'furring up', with deposits of lime scale. There is more detail about this in 'water heating systems' (see page 77).

fig. 38: kindling burning

Check the fire regularly while it is getting going and all the time it is alight to keep it burning well (see fig. 39).

fig. 39: stove lit and doors closed

clean air act

The Clean Air Act was introduced to reduce air pollution and it is a bit of a nightmare trying to interpret and apply it to using wood stoves. I will try and make some sense of it here but can't give you definitive guidance and it is well worth you looking on the Internet for information yourself.

I used the Office of Public Sector Information website for details, see 'resources' (page 141). The Act explains that anyone responsible for emitting 'dark smoke' could be liable for prosecution. Most of the act is specific to industry but the dark smoke also applies to domestic properties. There are exemptions to cover lighting of the stove when the fuel is not up to temperature, failure of some part of the stove and ancillaries, where correct fuel was not available and practical steps had been applied to reduce dark smoke from the inferior fuel.

Dark smoke is defined, in paragraph 3 section 3, but this is about as clear as dark smoke itself. It refers to the Ringelmann chart and states that dark smoke 'would appear to be as dark as or darker than shade 2 on the chart'. (The Ringelmann chart is a system used to define dark smoke when the Act was originally passed in 1956. It has five shades of grey with 0 being clear and 5 being black. Smoke is considered 'dark' if it has a shade of 2 or darker.) The Act also seems to say that actual proof of the smoke colour is not required.

The way I see it, and this is only my opinion, is that to prevent continuous emissions of dark smoke the following is required:

- your firewood must be well seasoned to below twenty per cent moisture content, as described elsewhere in this book
- you must use a combination of air inlet and chimney damper to keep a bright flame burning the gases. this is only possible with correct moisture content in the fuel. A Clearview stove will perform better than the older types of stove.
- don't shut the stove down tight for long periods as this will lead to the temperature falling and the gases will not be able to burn properly.
- smoke will always be produced when lighting the stove and when adding fuel, but this can be kept to a minimum by using appropriately-sized firewood.
- don't annoy the neighbours by cutting wood at inappropriate times. This will help to prevent people getting 'a bee in their bonnet' about things that they would otherwise never notice.

There are also smokeless zones that can be enforced under this Act and a quick call to your local council (environmental department) will find out if you are living in one. They usually apply to densely populated areas. If you live in one of these areas it may well be best to fit a pellet-fuelled stove where the burning process is fully automated and so emission control is tighter. There are other advantages to this type of stove in that the storage of fuel takes less space and there is no seasoning required. If there is some smoke emitted, then it is down to the technology or fuel and not you.

Having said all this, would you believe it that many people just fit wood stoves without taking notice of all the attendant bureaucracy? I'm not recommending this of course because it would be fateful to do so in print, but one must be careful of the clipboards whose self importance cannot countenance a deviation from the straight and narrow.

cooking with wood

The most basic type of cooking with wood is directly over the flames. The cooking ranges of the past had an open fire with an oven to one side. Trivets – hinged plates that could be swivelled over the fire – were used to support kettles and the like. The ovens commonly had flues running underneath them to divert extra heat from the fire to the oven. These could be closed when the oven was not in use by sliding a flat steel plate across the flue to block it off.

Things have moved on and the modern wood/solid fuel cooker is cleaner, more efficient and provides hot water. There are many types available and your choice of which one to buy will be determined by various factors: what is available, whether you want to buy new or second-hand, price, specific features that suit you, plus the 'modern' factors of colour, style and other things that don't matter to me.

I will not recommend specific makes here, particularly as these will mean little if you are living the other side of the world from me, but will mention brand names when referring to stoves I have used and their specific features. This is intended as an example of features to look out for, rather than a recommendation.

We have on old Rayburn number 3 in our kitchen (see fig. 40). This is a basic range with a long rectangular hot plate, small back boiler, and an oven that is heated from the side and top. There are no flue passages under the oven and so the oven takes quite some time to warm up. It was built as a coal stove and has a grate at the bottom with an ash pan underneath. The air inlet is in the ash pan door and so this range will burn wood well as long as some air passages are kept open through the ash build-up on the grate. To attain a good, roaring heat I often keep the top stove door, which is for adding fuel, just slightly ajar – about 5mm. This provides a secondary air inlet and keeps the gases burning well (see fig. 41).

fig. 40: Rayburn no. 3

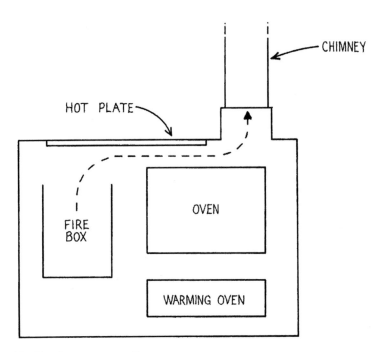

fig. 41: Rayburn gas path

The range in the Ecolodge has many modern and more efficient features. The oven has flues on both sides and underneath, and the boiler is of the wrap-around type. I have modified this range to give a more varied heating performance. Originally the chimney flue was attached at the bottom of the stove, below the oven. The flue gases had various paths around the oven to get to the chimney connection. This meant that the oven was always being heated and the gases being cooled. In summer the whole of the stove got hot, whereas all that was required was hot water and perhaps one hot plate.

The modification involved building a manifold that allowed gases to escape from either the base of the oven, as in the original design, or from the top of the oven just under the hotplate. A simple, chimney-damper-type valve blocks off the top exit when the oven requires heating (see fig. 42). This Tirolia range also has a grate that can be raised up the firebox, by using a simple T-bar tool. When the grate is at the bottom of the firebox, the entire wrap-around boiler is exposed to the fire. In this position it is easy to boil the water if a hot cooking fire is used. By raising the grate

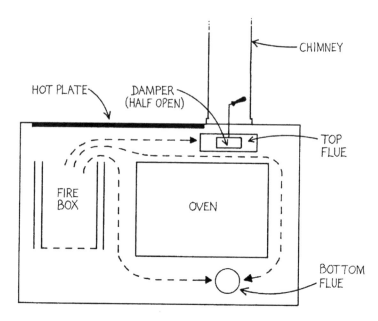

fig. 42: Tirolia gas path

a much smaller area of boiler is exposed to a roaring, cooking fire and the oven temperature can be kept high without the risk of the water boiling. It is quite clever stuff and it was not until I installed this second-hand stove, that I realised its full potential and flexibility (see fig. 43).

cooking heat

As with all stoves the size of the firewood determines the type of heat. With cooking this is very important, in that a quick, bright fire of small, split timber will get the hotplate or oven up to temperature quickly. Larger pieces are then used to keep the temperature up to that level. It is important to have smaller pieces readily available to boost the heat if necessary.

The hotplate has a cover that can be lifted up. On the Tirolia this completely covers the top of the stove, whereas on the Rayburn it just covers the hot plate. This cover keeps the heat in the hotplate and the oven. If you are not cooking or are just using the oven, keep the cover down. Using the cover is important: for example a kettle will boil quickly because of the stored up heat available. When using the hotplate another method of controlling

fig. 43: Tirolia Cooker

the heat of a specific pan is to move the pan to a cooler part of the hot plate, which is further away from the fire. It is just a matter of practice and learning to shuffle the pans about.

regulating oven temperature

The ovens in the Rayburn and the Tirolia both have a regulator plate that can be slid into position near the top of the oven. This reflects some of the heat and prevents localised heating, whilst keeping the rest of the oven at a regular heat. The Rayburn has a warming oven below the main oven. There is a small hole between the warming and main oven, and another hole between the main oven and the flue that goes over the oven and under the hot plate. These are there to allow cooling of the oven. The simple act of opening the warming oven allows cooler air to be drawn through the ovens and into the flue, by the suction of the chimney or draught. Beyond this direct form of heat regulation the fire can be allowed to die down, which takes time.

The chimney damper on the modified Tirolia can be opened to prevent flue gases from circulating around the oven, and so reduce heat transfer to the bottom and sides. The main principle is to bring the oven to temperature steadily and then reduce stove output to keep the temperature stable.

home stove building

In this chapter I am not going to describe how to build a stove from scratch. In the best tradition of 'down home' production, I have always built stoves from a base that started life as something else. These include a battery box from an electric forklift and a large diameter piece of tube. The inventive home mechanic will have the intuitive vision to see second uses for things, without being aware they are doing so. Is that a forklift battery box? No, it's a stove in the waiting.

What I do plan to do is to outline various techniques that I have used to make construction easier, and to avoid cost. If we were all tripping over excess wedges of cash, then the last thing we would do was to worry about building a wood stove.

Tools required for the home stove builder are a basic, well-equipped toolbox plus an arc welder, nine-inch (23cm) angle grinder, drill and PPE (personal protective equipment).

door and frame

The door and frame are critical. There must be a seal between them that can be easily replaced, and allow for shape changes within the stove caused by the heat. The most obvious type of change is caused by the fact that heat can warp the doors and, as a result, reduce the effectiveness of the seal. The doors I have made are to a basic design shown in the photographs (see fig. 46, 47 & 49). The door frame is often made of two-inch (5cm) angle iron. This is welded together so that one part of the angle iron faces forwards and the other faces outwards (see fig. 44)

The door is constructed around another angle iron frame. This time the door carcass is made to fit loosely over the doorframe, with the angles facing inwards as well as backwards (see fig. 45).

Describing this sort of thing is not easy and I'm sure reading it is even more confusing. What follows is a series of photos and line drawings. The basic idea is to provide an edge facing out from the stove. This edge fits into a wide groove in the door, into which is stuffed the seal material. I usually use fibreglass insulation rolled into a rope about 3cm in diameter. This is then stuffed into the groove and when the door shuts it beds in to match the frame. You can buy fire string, from a stove shop or hardware store, for this purpose. I don't know what fire string is made from now but it was originally made from asbestos, which means you must be very careful when dealing with, or using parts from, old stoves.

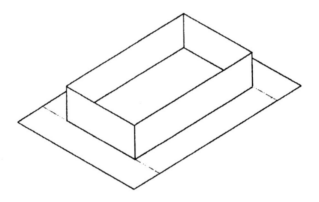

fig. 44: door frame

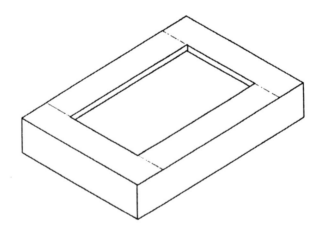

fig. 45: door carcass

fig. 46: doorframe, door and seal

The doorframe is made to a size that suits the stove body and is welded in place. The door carcass fits very loosely over it; I normally use two-inch (5cm) angle iron for this and weld a 3-4mm steel plate over the front to make the face of the door. Then I weld a piece of flat, steel strip, about 3mm by 30mm, on the inside to create the seal groove (see fig. 48).

fig. 47: door seal

This seal groove is about three centimetres wide, which gives plenty of leeway when fitting the door to its hinges.

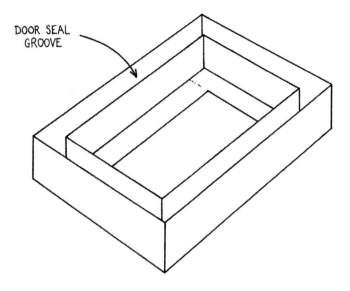

DOOR SEAL GROOVE

fig. 48: door seal groove

hinges

The hinges should be of the Parliament-type design. This means that the pivots are set back from the door and frame, so that the door can open wide and fold back if required. The reason for using this style of hinge is so that the door approaches the frame as squarely as possible. This enables the stove builder to obtain a good seal without having problems with the frame flange getting caught in the door-seal groove. If the hinge pivot point is too close to the door then it will be very difficult to attain effortless door action.

fig. 49: stove door and hinges

making a round stove

Making a stove from a large tube brings very specific construction problems: fitting a door to and attaining a good door seal is one of them. The answer to this is to fit a square frame into the side of the tube. The following photo (see fig. 50) shows the idea and you can see from looking at the base that is was made from an old gas cylinder. This is central to the spirit of using what is lying about to make a stove. I have also visited an agricultural engineering workshop where the stove was made from several lorry wheel rims stacked up and welded together.

fig. 50: round stove

air inlets

These have been described in detail in other chapters (see page 57 in particular). Their main function is to allow air into the two important parts of the fire: to the base of the fire, to create turbulence and provide oxygen to the charcoal, and to the flammable gases within the heated zone. The common construction is a plate sliding over a hole to adjust the volume of air (see fig. 49). Some stoves are fitted with an extra, thermostatic air inlet. These are built using bi-metal-strip technology, and they close as the temperature rises. They are adjustable and can be useful to keep the stove burning efficiently and as cleanly as possible. Some older, coal-burning room-heaters, for example Rayburns, were fitted with these thermostatic air inlets, and so, if you can find a scrap one, could be a source for the stove builder.

building regulations

The building regulations for fitting a stove are to be found the government's Planning Portal website, see 'resources' (page 141) under document J (Combustion appliances and fuel storage systems). There are appendices to the main documents with other contacts with further, detailed information, see 'resources' (page 141). There is some basic information for the wood stove installer within document J, but you have to wade through a lot of other information to get to it. The following is relevant though.

The hearth should be at least 840mm square and project at least 500mm into the room. The surface of this non-combustible area should be 125mm thick and project at least 300mm forward from the stove. There should be a gap of at least 50mm between the underside or edge and any combustible material. Between the stove and any heat resistant wall there should be a minimum of 150mm gap.

Beyond 150mm it seems, from document J diagram 30, page 36, that no protection of combustible material is required, which seems a bit odd to me, and I would suggest a gap of 500mm would seem safer, although that is only my opinion. From personal experience I know that stoves radiate considerable quantities of heat and to have, for example, a pine wall with the resin melting out of the knots just 150mm away would not be a safe idea.

sharp bits

The history of industrialisation is littered with stories of people losing bits of themselves. 'Sharp bits' and inattentiveness do not go together. A friend of mine nearly lost a finger while splitting kindling. All but a bit of tendon was severed, followed by weeks of pain, splints and, finally, he was left with a lack of feeling in the sewn-on finger. None of us want to have an amputation for the sake of wood fuel and being self-sufficient, so in this chapter I am going to give some basic ideas about safety and to get everyone thinking about safe working practice.

To bring my point home please be aware that sharp-edged tools cut deeply and with no warning, circular saws cut bits off and remove flesh to the width of the blade and chainsaws lacerate, remove flesh and bone, rip and tear, and introduce oil and muck deep into the wound. Cutting a major artery can lead rapidly to premature death.

personal protective equipment

Otherwise known as PPE, this is a must in many working situations. The level of PPE depends, of course, on the work you are doing and the associated hazard and risk. The main criterion is that the PPE should be suitable for the task (see fig. 51).

There should be:

- gloves to protect the hands
- steel toe-capped boots for the feet
- safety glasses for the eyes
- full chainsaw kit, when using a chainsaw

Chainsaw PPE consists of chainsaw boots, trousers, gloves, and forestry helmet. All of these, except the forestry helmet, are marked with the CE logo, see 'resources' (page 141) of a chainsaw in a shield.

The boots have steel toecaps, are high to protect the ankle, and contain layers of fibre that the chain will not cut easily. If cut, the fibre is pulled out of the boot by the cutters and clogs the chain and sprocket abruptly.

fig. 51: chainsaw PPE

Gloves have a protective pad of chain-clogging fibre on the rear of the left hand. The left hand is always on the front handle and so is closer to the chain.

Trousers have the same layers of clogging fibre and come in several styles. The main thing is that they are either front protection (type A), or all-round protection (type C). All-round protection is what is officially required, but they are hot to wear in summer. There are also various grades of protection depending on chainsaw speed

Class 1: 20 metres per second
Class 2: 24 metres per second
Class 3: 28 metres per second

In forestry we use class 1 trousers because they are lighter and cooler in summer. As with all PPE, the protective equipment is not a substitute for good working practice and training, but it is there as a last resort when unforeseen things happen.

Forestry helmets give protection to the head and keep the rain off. The ear defenders are a must for chainsaw use, and the mesh visor protects the face and eyes. In windy situations sometimes it is a good idea to wear safety glasses as well as the visor to keep fine dust out of the eyes. We also use these glasses when teaching chainsaw carving as carving produces both dust and chippings that readily blow everywhere.

hand tools

The main hand tools for processing firewood will probably be:

- large hatchet or small axe for splitting kindling
- splitting axe
- hand saw or bow saw
- sledge hammer and steel wedges

Most of these tools have been explained in 'buying and storing wood' (see page 21), but I feel some extra information regarding safety is required.

small axe

The safety issue here is fingers damaged by bouncing axes. To avoid this only use easy-to-split wood for kindling, so that the use of a large amount of force is unnecessary. I always put suitable kindling to one side as the yearly firewood stack is being used;

that way it's easy to find and it doesn't get used as standard firewood.

fig. 52: preparing kindling (1)

The small axe I use for splitting kindling (see fig. 52) has a relatively large head and a short shaft. The reason for this is so that you use the weight of the head to do the splitting rather than speed and momentum. I hold the shaft about three inches from the head and don't move it very far. This is the best method to use when you are holding the block of potential kindling with your other hand. The idea is to avoid directional inaccuracies when swinging the axe from any distance.

Another method that achieves the same level of finger safety is to place the axe edge on the log and then move both log and axe together to strike the chopping block from a distance of about six inches (see fig. 53). The axe does not need to have a super-honed, razor edge; a ground finish from a whetstone wheel, or the coarse side of an oilstone is sufficient. The idea is that the edge should be sufficiently sharp to penetrate the end grain fibres of the wood, and from that point the shape of the axe head splits the kindling.

fig. 53: preparing kindling (2)

splitting axe and hand saw

In 'buying and storing wood' (see page 21) I highlighted the oiling of saw blades as important to reduce the loss of a fine edge to rust. A felling axe should never be used for splitting wood as it will get stuck easily, due to its shape, and the shaft is not as robust as on a splitting axe. A glancing blow from a felling axe to legs and feet is extremely dangerous, and these axes should not be used for splitting or the removal of branches.

sledge hammer and wedges

A good, medium-weight sledgehammer is a very useful tool, for a variety of tasks. Used in conjunction with steel wedges it is invaluable for splitting knotty timber.

fig. 54: sledge and wedge

The sledge should be of the correct weight for the operator; there is no point in struggling with one that is too heavy to be used effectively. Always have a secure footing at the chopping block, make sure the block is the right height, and don't ask anyone to hold the wedge. The wedges should be struck square on with a resounding blow (see fig. 54). Striking the wedge at an angle will cause the sledge to glance off at an angle and it could be wrenched from the operator's hands. The wedges are made of steel to withstand the constant blows but with use the tops tend to burr over. These burrs continue to grow until they break and fly off and can be very dangerous – particularly to the eyes. As burrs develop they should be ground back to a slight taper, which will help to prevent their development.

saw bench

A circular saw should be running with a peripheral speed of 3000 metres per minute to cut most efficiently. This means there is a lot of energy stored in the rotation of the blade. This energy will cause logs to roll into the blade when the cut is started and then kick, which can bring unwary hands in contact with large, pointy teeth. I keep the belt drive a bit slack on my saw so it can act like a slip clutch. As with all machinery, training in machine maintenance and good working practice is of extreme benefit.

chainsaw

I am including some basic information on the safety aspects of using chainsaws but want to emphasise that this should not be seen as a substitute for proper training. Lantra Awards, the governing body for chainsaw training, has agreed to provide images from their excellent workbooks, which are provided when you go on a Lantra registered chainsaw course, see 'resources' page 141, for this section.

There are two basic power systems for chainsaws: electric and petrol.

electric chainsaws

Electric saws are plugged into the mains electricity supply through an RCD protection unit, which commonly disconnects the power in 30 milliseconds if there is a fault or leak to earth, etc. Unlike the petrol saws there is no variable speed, no noise, and when the saw is put down it is off, not ticking over. So there are plenty of benefits to an electric saw but it cannot be used away from mains power without a generator, which is severely limiting. One other major disadvantage is that, because the saw doesn't regularly run out of fuel, operators are tempted to keep going without breaks. This can lead to fatigue and vibration problems, for example vibration white finger, but also the chain oil will run out and so damage to the drive and chain will occur. Please see page 134 for more about oils.

petrol chainsaws

These saws use a two-stroke engine, which runs on petrol that has had specialist oil added. The oil lubricates the working parts of the engine and prevents seizure. They are noisy, a bit smelly, very mobile and the modern professional models are built to low-vibration standards.

chainsaw safety features

There are ten recognised safety features on a chainsaw (see fig. 55):

- positive and clearly marked on/off switch
- chain brake
- eye/ear defender symbols
- safety throttle with 'dead man's' handle
- chain catcher
- rear-handle extension chain guard
- anti-vibration system
- exhaust pointing away from operator
- safety feature on bar and chain combination
- bar cover

positive and clearly marked on/off switch
The switch should have a positive action, so when it is off you know it's off. It needs to be clearly marked so that if there is a problem anyone else can see it and can switch it off. If the switch fails to switch off, operate the choke to stop the saw and do not use it until the switch is fixed.

chain brake
This stops the chain in the event of kickback and also allows the operator to prevent any accidental chain movement by operating the brake before taking either hand off the saw, or before walking more than several paces with the saw running.

eye/ear defender symbols
These are to remind you to use the right PPE.

safety throttle with 'dead-man's' handle
This is to prevent any accidental operation of the throttle by branches and other pokey-type things.

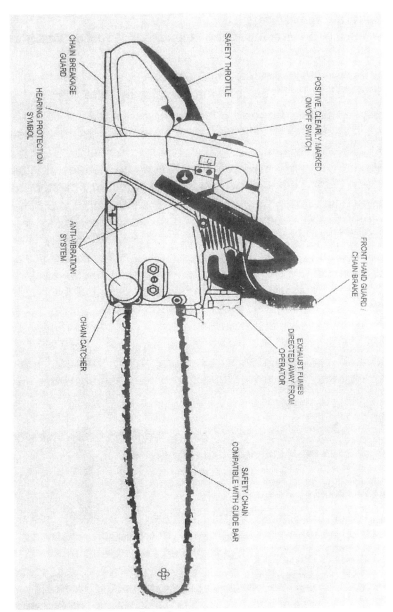

fig. 55: chainsaw safety features

chain catcher
This will catch the chain in case of derailment or chain breakage.

rear handle extension chain guard
This protects the right-hand knuckles from derailed or broken chain.

anti-vibration system

This should prevent operator fatigue and vibration-white-finger circulation problems.

exhaust pointing away from operator

To avoid excess poisoning of the operator by petrol fumes.

safety feature on bar and chain combination

This operates as part of the extended depth-gauge and guard-link anti-kickback system. If you look on the chainsaw cutter illustration (fig. 57) you will see the depth gauge situated in advance of the cutting edge. This regulates the thickness of wood that each cutter can remove in any one pass. There is a slope to the front of the depth gauge which reduces the ferocity of any potential kickback to some extent. The slope effectively lifts the bar and chain slightly out of the cut as it reaches the kickback zone – the top quadrant of the bar. Remember that the chain runs forward on the top of the bar, and as it goes round the nose the cutter has a tendency to grab too much timber. The guard links seen in the saw-chain parts illustration (fig. 56) extends the slope of the depth gauge and works to the same end. Saw chains for occasional-user chainsaws have huge guard links to fill in the gaps between cutters to reduce to a bare minimum the risk of kickback.

bar cover

This prevents damage to the cutting edge and operator's body when the saw is in transit and not being used.

fuel and oil

Fuel for a two-stroke, petrol chainsaw is basically standard, unleaded petrol with two-stroke, lubricating oil added by the operator. Five litres are usually mixed at a time at a ratio of fifty parts fuel to one part oil; the 50:1 oil has been specially developed to provide correct lubricating qualities combined with low smoke emissions. It is best to use a recommended manufacturer's product: professional quality products are available from manufacturers like Stihl, Husqvarna and Oregon. At 50:1 it's not easy to measure without experience, and so one-shot sachets are available.

chain oil

This has a separate tank and in all but the nastiest saws has a mechanically-driven pump. The oil has a high viscosity so that it sticks to the important parts of the bar and chain. The oil is pumped through a hole in the side of the bar and into the bottom of the groove, where it is dragged round by the drive link to lubricate the working parts.

bar and chain

The bar holds the chain in place and enables the operator to direct the cut. There is a groove on the edge of the bar in which the drive link of the chain runs. This means that, when the chain is tensioned correctly, the chain is unable to come off the bar. The chain has cutters on the top, which cut the wood. These need to be sharpened routinely with a round, saw file of the correct diameter for the type of chain fitted. Saw files can vary between 4mm to 5.5mm depending on chain type – there are quite a few. The file is held in a filing guide specific to the size of the file, and the cutters are sharpened at the correct angle for that type of chain. Depth gauges set the depth of cut and act as an anti-kickback feature (see fig. 56 & 57). Instruction on sharpening the chain is included in all good training courses.

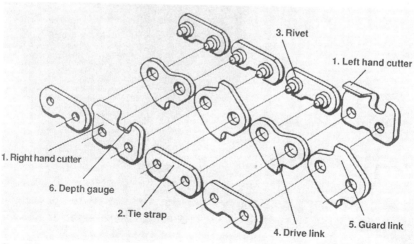

fig. 56: saw chain parts

kickback

The top quadrant of the bar is a place where the saw chain is running forwards and bending downwards to go round the nose. This is called the 'kickback zone' and should not be used for cutting by anyone except an experienced professional. At this point there is a tendency for the cutter to stand proud of the chain. If this part of the chain comes into contact with anything it can have the effect of throwing the saw back at the operator, with potentially disastrous results. This feature is one more reason to get good-quality training through a registered course.

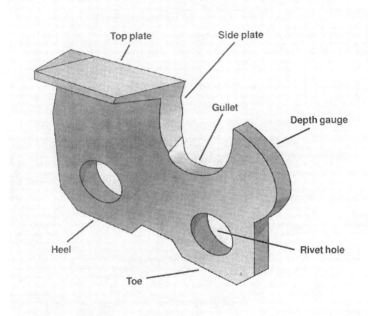

fig. 57: chainsaw cutter

chain brake

This is in front of top handle (see fig. 55), prevents the chain from moving when engaged and should be used at all times except when you are cutting. The chain brake can also come on during kickback so that the saw kicks back and the bar flies up in an arc backwards towards the operator. The chain brake lever comes into contact with the back of the operator's left hand and puts the brake on, stopping the chain.

viewpoint

This is a general rounding-up of wider issues that, as I write, are becoming increasingly important in the public arena. So, you have installed or are thinking of installing a wood heating system. The decision-making process for becoming a wood-heat system owner is not that complex and is usually based on a combination of economics and concern about reducing our carbon footprint. This is fair enough and these two work to some extent hand-in-hand. However, there are other motivations that include designer-appeal, marketing forces, style, and thinking that 'it's the latest thing to own'. Now correct me if I'm wrong, but if a heating system is fitted for the latter reasons, then its doubtful that it will be used very often. This represents a huge waste of resources and energy.

We are all responsible for our own carbon footprint and to resist the aggressive tide of pressure to consume. The production machine needs to sell more and more to perpetuate itself and in so doing uses up more land, energy, and materials.

So, let's get to the crux of the matter: if you are fitting a wood burner because it's the current thing to do, then think again, and then think about what you are doing to yourself, the country, and the world by being a dispassionate and unthinking consumer. At the very least you are at risk of being taken for a ride by clever marketing and false dreams of liberation.

pitfalls

Low-carbon living is about making choices based on your personal preference and its impact locally and globally. If low-carbon living is aligned with a reduction in personal needs then it can also give rise to more free time. Divorce yourself from the pressures of consumer society and there is less desire to spend hard-earned cash. In my opinion it's always better to earn less and spend less and have more time to just mess about with things and learn in the process.

A wood-burning heating system allows you to reduce bills for energy and contributes to a carbon-neutral existence. All this benefit can, however, be negated by a holiday flight, or huge commuting miles to work. This is where low-carbon living extends into the rest of your life and, as you get older, you realise that time is more important than wedges of cash and stress.

I am including something that my friend Bob Miller, who has been using a wood burner for a long time, has written to illustrate what I mean.

'Part of the appeal of a wood-burning stove, especially to a cash-careful person like me, is that some of the fuel for it is free. We installed our stove twenty-five years ago and have never regretted it.

There are some disadvantages with wood burning; the greatest in my opinion is storing so much fuel and the space it takes up. You really need to store through one winter the fuel that you will need for the next. You can't necessarily trust a wood merchant when he says "it's nice and dry", and dry is the important factor.

The burning process of not-totally-dry wood produces a sticky smoke, which contains chemicals that are not good for chimney or stove. So, have your chimney lined and insulated and run the stove to obtain a bright flame rather than letting it smoulder. We had to replace our first boiler because we didn't do this and corrosion got the better of the thin boiler steel.

The benefits gained from burning wood far outweigh these disadvantages, and one of the best is that from time to time you will be presented with free fuel from members of the unthinking, consumer society. You will never get this from the utility companies and, what's more, you don't have to depend on some unreliable boiler engineer putting unnecessary parts in your boiler and then charging a fortune just so they can have yet another holiday abroad or a new car or a new nonsense telly.

You need to be a bit careful with this free fuel to make sure it does not include treated timber, like telephone poles and fence posts, or chipboard and melamine facings, all of which give off poisons when burnt.'

So there you are, a slightly different view from Bob. But being part of a wood-heat household becomes a way of life and, in my opinion, a large stock of firewood that would last the next five years is a buffer against possible mean times in the future. You will at least be able to keep warm and cook you own, home-grown food.

resources

wood stove suppliers / installers

Here's just a small selection of the many suppliers of wood stoves around the country. Try Yellow Pages and Google too.

Logpile
www.nef.org.uk/logpile
01908 665555
Database of wood stoves/boilers and wood pellet stoves/boilers.

Stoveland
www.stoveland.co.uk
Fantastic online resource – listings of stove and flue manufacturers and suppliers.

Canvas & Cast
Rosemount, Canada Hill
Ogwell
Devon TQ12 6AF
www.canvasandcast.com
01626 363507
Small cast iron wood stoves.

Clearview Stoves
More Works
Bishops Castle
Shropshire SY9 5HH
www.clearviewstoves.com
01588 650401

Fires Online
Station Yard
Chester Street
St. Asaph
Denbighshire LL17 0RE
www.firesonline.co.uk
0845 00 44 333

Stoves Online
Capton
Dartmouth

Devon TQ6 0JE
www.stovesonline.co.uk
0845 226 5754
Stoves, flues, plus loads of useful information.

Windy Smithy
Bishops Plot
Blackborough
Cullompton
Devon EX15 2HY
www.windysmithy.co.uk
07866 241783
Handmade stoves – including beautiful little stoves for yurts or camper vans.

Woodburning Stoves Ltd.
Waterlands
Fenwick Road
Stewarton
Ayrshire KA3 5JE
www.woodburningstoves-ltd.com
01560 483966
Stoves and stove products; family business.

HETAS
Orchard Business Centre
Stoke Orchard
Gloucestershire GL52 7RZ
www.hetas.co.uk
0845 634 5626
Trade body for solid fuel domestic heating appliances, fuels and services; website includes list of solid fuel appliance installers and chimney sweeps.

Salvo
www.salvo.co.uk/directory.html
Try this site for second-hand stoves – salvage yards.

Louth Tractors
Tattersall Way, Louth
Lincolnshire LN11 0HH
01507 605441
www.louthtractors.co.uk
Wood chip/pellet boilers

firewood / kindling suppliers

Logpile
www.nef.org.uk/logpile
01908 665555
Information; plus database of UK wood (and kindling) suppliers, plus wood pellet suppliers.

Ecolots
www.ecolots.co.uk
List of wood-fuel suppliers around the country.

Woodlots
The Woodland Enterprise Centre
Hastings Road
Flimwell
East Sussex TN5 7PR
www.woodnet.org.uk/woodlots
01580 879552
List of wood-fuel suppliers in the south-east of England.

Bingley Logs
26 Heaton Drive
Eldwick, Bingley
West Yorkshire BD16 3DN
www.bingleylogs.co.uk
01274 566997
Deliver firewood all over the north of England.

Four Seasons Fuel
Four Seasons Farm
Coneyhurst, nr Billingshurst
West Sussex RHI4 9DG
www.fourseasonsfuel.co.uk
01403 783379
Family firm supplying logs, kindling and charcoal.

Eco-Logs
Tyn-y-Bryn
Deriside, Abergavenny
Monmouthshire NP7 7HT
www.eco-logs.co.uk
01873 856682
Briquettes made from compressed sawdust and shavings.

courses / training

LILI
Redfield
Buckingham Rd, Winslow
Bucks MK18 3LZ
www.lowimpact.org
01296 714184
Courses: heating with wood / how to make a wood burner from a gas bottle / managing small woodlands.

Centre for Alternative Technology
Machynlleth,
Powys, SY20 9AZ
www.cat.org.uk
01654 705950
Heating with wood course.

Woodland Training Division
International business School
Rose Court, Station Road
Old Leake, Boston
Lincs PE22 9RF
www.internationalbusinessschool.net
01205 870062
Forestry & chainsaw training.

Lantra Awards
Lantra House
Stoneleigh Park, nr Coventry
Warwickshire CV8 2LG
www.lantra-awards.co.uk
02476 411655
Governing body for land-based training.

National Proficiency Test Committee
Stoneleigh Park, nr Coventry
Warwickshire CV8 2LG
www.nptc.org.uk
02467 857300
The largest nationally-recognised awarding body within the land-based sector.

Training Pages
www.trainingpages.com/x/category,kw-1422,.html
National listing of chainsaw trainers.

books

All these books are available via LILI's website:
www.lowimpact.org

The New Woodburner's Handbook: A Guide to Safe, Healthy and Efficient Woodburning
Stephen Bushway
Storey Publishing, £9.99
US book – guide to safe, healthy and efficient wood burning.

Natural Home Heating
Greg Pahl
Chelsea Green, £22.50
US book – sections on wood stoves, masonry stoves, solar heating, geothermal and heat pumps.

The Book of Masonry Stoves: Rediscovering an Old Way of Warming
David Lyle
Chelsea Green, £35.00
The first comprehensive survey ever published of all the major types of masonry heating systems, ancient and modern.

The Woodburner's Companion: Practical Ways of Heating with Wood
Dirk Thomas
Alan C Hood & Co. £12.95
Another US publication - guide to the ins and outs of the many ways available to burn wood for heat.

The Ax Book: the Lore and Science of the Woodcutter
Dudley Cook
Alan C Hood & Co. £16.99
Ideal resource for anyone who wishes to fell trees and take lumber or firewood from the forest.

The Forgotten Art of Building a Good Fireplace
Vrest Orton
Alan C Hood & Co. £6.95
Basic principles of fireplace design with the help of drawings.

Choosing and Using Flues & Chimneys for Domestic Solid Fuel and Wood Burning Appliances
British Flue & Chimney Manufacturers Association
Free download. From LILI's website or below:
http://www.feta.co.uk/downloads/guide-yellow1.pdf

Woodlands: a Practical Handbook
Elizabeth Agate
BTCV £13.95
Covers the full range of woodland work, from tree planting and establishment, through to thinning and conversion into woodland products.

The Woodland Way: a Permaculture Approach to Sustainable Woodland Management
Ben Law
Permanent Publications £16.95
Practical alternative to conventional woodland management.

Woodland Management: a Practical Guide
Chris Starr
Crowood Press £16.99
Buying and owning woodland – what to look for and how to avoid the common pitfalls.

Tree Planting and Aftercare: a Practical Handbook
Elizabeth Agate
BTCV £14.95
Comprehensive guide to establishing native trees.

Plumbing: Heating and Gas Installations
R D Treloar
Blackwell £19.99
Best plumbing book we've found – for plumbing in your back boiler and solar connections.

information / associations

Solid Fuel Association
7 Swanwick Court
Alfreton
Derbyshire DE55 7AS
www.solidfuel.co.uk
0845 601 4406

Encouraging greater awareness of the benefits of domestic solid fuel heating; unfortunately includes coal.

Villager Stoves
Millwey Industrial Estate
Axminster
Devon EX13 5HU
www.villager.co.uk/rightstove.htm
01297 35596
Online calculator to find the size of stove for your needs.

Euroheat
Unit 2, Court Farm Business Park
Bishops Frome
Worcestershire WR6 5AY
www.euroheat.co.uk/heating_capacity.htm
01885 491100
More information about the size of stove required for your living space.

Office of Public Sector Information
www.opsi.gov.uk/ACTS/acts1993/ukpga_19930011_en_1
The Clean Air Act 1993

Planning Portal
www.planningportal.gov.uk/england/professionals/en/400000000 0503.html
Building Regulations covering combustion and heat-producing appliances.

National Association of Chimney Sweeps
Unit 15, Emerald Way
Stone Business Park, Stone
Staffs ST15 0SR
www.chimneyworks.co.uk
01785 811732
Information plus listing of UK chimney sweeps.

Conformance
Great Hucklow, Buxton
Derbyshire SK17 8RG
www.conformance.co.uk/directives/ce_ppe.php
01298 873800
Personal Protective Equipment Directive.

Health & Safety Executive
www.googlesyndicatedsearch.com/u/HSEC?q=chainsaw&sa=Go
0845 345 0055
Lots of information on chainsaw safety.

forestry / trees

Forestry Journal
PO Box 7570
Dumfries
Dumfriesshire DG2 8YD
www.forestryjournal.co.uk
01387 880359
Articles, forestry equipment for sale.

Royal Forestry Society
102 High Street
Tring
Hertfordshire HP23 4AF
www.rfs.org.uk
01442 822028
Lots of info; increasing people's understanding of forestry.

Forestry Commission
231 Corstorphine Road
Edinburgh
EH12 7AT
www.forestry.gov.uk
0845 367 3787
Promoting forests for ecology, education, leisure and employment.

Forestry Contracting Association
Tigh na Creag
Invershin, Lairg
Sutherland IV27 4ET
www.fcauk.com
0870 042 7999
Protecting the interests of forestry workers.

Small Woods Association
Green Wood Centre
Station Road
Coalbrookdale

Telford
Shropshire TF8 7DR
www.smallwoods.org.uk
01952 432769
Supporting the owners of small woodlands; index of UK groups involved with woods and woodland skills.

Arboricultural Association
Ampfield House
Romsey
Hampshire SO51 9PA
www.trees.org.uk
01794 368717
Information and advice about the planting and care of trees for the general public.

British-Trees
www.british-trees.com
Definitive online guide to native British trees.

British Horse Loggers
Hill Farm, Stanley Hill
Bosbury, Ledbury
Herefordshire HR8 1HE
www.britishhorseloggers.org
01531 640236
Association of contractors extracting felled logs from woodlands using horses; register, newsletter, training, for sale & wanted, etc.

tree nurseries

Click Forestry
Silvicultural Systems Ltd
Stour Valley Business Centre
Brundon Lane, Sudbury
Suffolk CO10 7GB
www.click-forestry-directory.co.uk
01778 440716
Listing of UK tree nurseries

Ecolots
www.ecolots.co.uk
Another listing of tree and plant nurseries.

Aveland Trees
Dunsby
Bourne
Lincolnshire PE10 0UB
www.avelandtrees.co.uk
01778 440716

British Hardwood Nurseries
Norton Road
Snitterby, nr Gainsborough
Lincolnshire DN21 4TZ
www.britishhardwood.co.uk
01673 818443

British Trees & Shrubs
125 Hansford Square
Combe Down
Bath BA2 5LL
www.britishtrees.co.uk
01225 840080

Cheviot Trees
Newton Brae
Foulden
Berwick upon Tweed TD15 1UL
www.cheviot-trees.co.uk
01289 386755

Chew Valley Trees
Winford Road
Chew Magna
Bristol BS40 8HJ
www.chewvalleytrees.co.uk
01275 333752

Maelor Forest Nurseries
Fields Farm
Bronington, Whitchurch
Shropshire SY13 3HZ
www.maelor.co.uk
01948 710606

Taynuilt Trees
Keepers Cottage

Taynuilt
Argyll PA35 1HY
www.taynuilttrees.co.uk
01866 822591

Thorpe Trees
Thorpe Underwood
Ouseburn
York YO26 9TA
www.thorpetrees.com
01423 330977

other

Ecolodge
Rose Court, Station Road
Old Leake, Boston
Lincs PE22 9RF
www.internationalbusinessschool.net/eco-lodge.htm
01205 870062
If you want to meet the author, go and stay in the Ecolodge, which he built with local timber, along with wind turbine, solar electric and hot water panels, compost toilet and, of course, wood stove / range.

Low Carbon Buildings Programme
www.lowcarbonbuildings.org.uk
08704 23 23 13
At the time of writing, providing government grants for automated wood pellet stoves and wood-fuelled boiler systems.

notes

Printed in the United Kingdom by
Lightning Source UK Ltd., Milton Keynes
139919UK00001B/5/P